Lecture Notes in Logistics

More information about this series at http://www.springer.com/series/11220

Matthias Klumpp · Caroline Ruiner
Editors

Digital Supply Chains and the Human Factor

Springer

Editors
Matthias Klumpp
Chair for Production and Logistics
Georg-August-University of Göttingen
Göttingen, Germany

Caroline Ruiner
Chair for Sociology
Institute of Education, Work and Society
University of Hohenheim
Stuttgart, Germany

ISSN 2194-8917 ISSN 2194-8925 (electronic)
Lecture Notes in Logistics
ISBN 978-3-030-58429-0 ISBN 978-3-030-58430-6 (eBook)
https://doi.org/10.1007/978-3-030-58430-6

This Springer imprint is published by the registered company Springer Nature Switzerland AG
The registered company address is: Gewerbestrasse 11, 6330 Cham, Switzerland

Contents

About the Editors

Matthias Klumpp is working at the Georg-August-University of Göttingen, FOM University of Applied Sciences Essen, and Fraunhofer Institute for Material Flow and Logistics (IML) Dortmund. He received his Ph.D. from the University of Leipzig, his habilitation from the University of Duisburg-Essen and was in visiting positions at University of Twente, EUI Florence, University of Lugano (IDSIA), and University of Modena. Addressed research topics are sustainable, digital (AI), and resilient supply chains with a focus on HCI and qualification.

Caroline Ruiner is professor for sociology at the University of Hohenheim in Germany. She studied sociology and management science at Goethe University Frankfurt, received her doctorate in sociology at the University of Augsburg, and habilitated at the Faculty of Social Sciences at Ruhr University Bochum. Her research focuses on new work and new employment relationships and the effects on individual and organizational levels as well as with regard to industrial relations.

Digital Supply Chains and the Human Factor—A Structured Synopsis

Matthias Klumpp and Caroline Ruiner

Abstract Digital developments and changes in the production, supply chain and logistics sector as well as specific concepts like automation and Industry 4.0 or the Internet of Things are omnipresent. Especially the human role in such settings experiences important changes, which has not been adequately addressed in research yet. This introduction chapter contains an overview of elements encountered in digitalization processes in order to ensure sustainable work environments and efficient Human-Computer Interaction settings for the benefit of workers and organizations. Thus, it is the aim of this chapter to provide a structured synopsis to consider the human factor in analyzing digital work processes. This synopsis is aligned with typical workflow developments in digitalization projects and can be transferred to different work settings in supply chains. Finally, we outline the chapter structure of this book within four thematic sections in order to provide a joint storyline on investigating the human factor in digital supply chains.

Keywords Human factor · Human-Computer Interaction · Digitalization · Work processes

1 Introduction

Digital developments in production, logistics and supply chain management are a major field of interest to management science research and business practice: A multitude of publications and concepts is dealing with this and has spurred further

M. Klumpp (✉)
Georg-August-University of Göttingen, Göttingen, Germany
e-mail: matthias.klumpp@uni-goettingen.de

Fraunhofer IML, Dortmund, Germany

FOM University of Applied Sciences, Essen, Germany

C. Ruiner
University of Hohenheim, Stuttgart, Germany
e-mail: caroline.ruiner@uni-hohenheim.de

M. Klumpp and C. Ruiner (eds.), *Digital Supply Chains and the Human Factor*, Lecture Notes in Logistics,
https://doi.org/10.1007/978-3-030-58430-6_1

academic work since the first discussions of concepts like Industry 4.0, Internet of Things or Physical Internet [10, 25, 35, 45, 58].

There are several research publications dealing with the phenomenon of digitalization in different industries such as in knowledge-intensive fields, in gig-work and other services [18, 20, 23, 43, 59]. Digitalization expectations are also linked to sustainability improvements as reduction of resource use and emissions are connected to optimization objective with AI implementation e.g. in the reduction of transport route distances [28, 37, 39, 60]. However, there is a lack of dedicated analysis regarding the implications of digitalization for the human workforce in the supply chain field. A serious amount of research focuses on operational questions of Human-Computer Interaction (HCI), but this is not comprehensive at the sight of groundbreaking changes for workers in logistics and supply chain management [8, 32, 34]. Traditional HCI concepts are focused on the communication and interaction interface of one robot or digital device with one human operating it. In future developments, the human factor has to be evaluated and aligned with digital devices on a more strategic and comprehensive level: Data management, interaction points and supervision as well as transparency questions have to be answered for a larger range of processes and workers at the same.

This book aims at closing this gap further as digital developments are increasingly shaping supply chain work environments in transportation, warehousing, picking and packing as well as production. Therefore, a profound analysis is required in order to ensure sustainable work environments and efficient HCI settings for the benefit of workers and organizations. It is the aim of this chapter to provide a structured synopsis to consider the human factor in analyzing digital work processes. Only if human workers are accepting the digital components introduced and are able to work effectively with them, a long-term social work relationship will evolve and therefore enable sustainable work engagements.

This chapter is organized as follows: Section 2 describes core elements when analyzing the human factor in digital supply chains as a structured synopsis. Section 3 outlines the structure of this book and the contributions of the individual chapters towards this challenge in describing and designing the human factor in digital supply chain processes.

2 Synopsis of Core Elements

In order to gain competitive advantages, organizations rely on the digitalization of value-added processes. There are new opportunities to digitalize work processes using mobile devices, handheld scanners, AI applications or smartphone apps. However, to achieve positive effects, the human factor has to be considered as has been outlined by research for a long time [7, 13]. Thus, we take into account the individual, social, physical, cognitive and organizational aspects of work [57]. There is a strong need to understand workers' attitudes, their expectations, motivations, work engagements and cooperation in socio-technical systems. A socio-technical

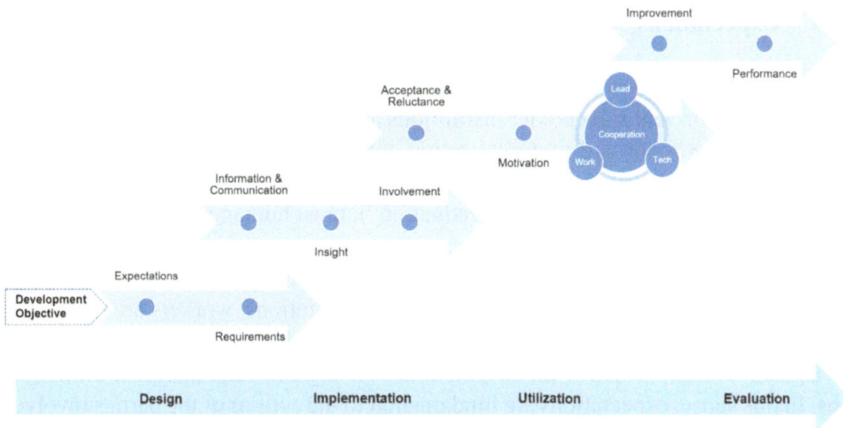

Fig. 1 Structured synopsis of the human factor in digital work processes

approach means that the human factor comes into focus and with this the social values of using new technologies at work [52].

The structured synopsis is oriented at the chronology of digitalizing work. In the first place, there is a concrete need to improve work processes and a development objective to be identified. Then, new technologies have to be designed to fill the needs defined (*Design Phase*). In a next step, new technologies will be implemented in specific supply chain processes (*Implementation Phase*). Thereafter, the day-to-day use in supply chain business contexts will follow (*Utilization Phase*). Finally, an evaluation estimating the quality and quantitative impact of the change will be conducted (*Evaluation Phase*). Figure 1 shows the development and its components and the subsequent sections outline the specific content elements of the phases and elements.

2.1 Design

The concept or design phase is about the design of new technologies that are needed to meet a development objective. However, this stage is not only about meeting technical requirements but also about how to design interaction between humans and machines as well as the design of the user interface [19]. Already Zuboff [61] emphasized that information technology can be designed with different intentions and modes of cooperation with humans (automatization vs. information). Each design has different consequences for workers such as controlling and deskilling or empowering and upskilling [44]. In this sense, it is relevant to consider workers' expectations and to direct attention towards the requirements of new technologies to improve working conditions and work processes.

2.1.1 Expectations

The human angle regarding digital supply chains starts with the question of expectations of humans and stakeholder institutions alike. Usually, for example efficiency expectations are connected to digitalization developments and projects. This could severely distort the further implementation and acceptance process as with an efficiency improvement objective ("cost reduction"), most human aspects up to job loss fears and realizations are perceived in a negative perspective and as technological determinism [1, 16, 41]. If on the other hand digitalization projects are initiated with quality improvement and customer innovation expectations, workers are often integrated and motivated in a positive fashion as the main objective is not cost reduction but a positive improvement with the potential of rising sales and therefore even more jobs. In this sense, expectations are fundamental to the actions of the parties involved [50]. Consequently, it is expected that well-being (e.g. health and safety, satisfaction, pleasure, learning, personal development) and performance (e.g. productivity, efficiency, effectiveness, quality, innovativeness) are intertwined [13]. Against this background, it makes sense to consider improving the workers' well-being through new technologies [5, 14, 30]. For digitalization concepts and projects, therefore a clear analysis of expectations as well as a professional expectation management are crucial success factors.

2.1.2 Requirements

Also, requirements for the use of new technologies must be explored at the beginning of the process [19]. This refers to the analysis of what is required to enable good digital work. This implies for example that technologies should not interrupt or complicate related work processes but rather provide support for workers. Additionally, it also needs to be clarified what requirements workers must have to deal with new technologies [5]. Thus, the necessary human capabilities and competencies have to be identified. Such a human-centered design approach anticipates the human perspective and is not only designed around technological requirements [42]. Rather, the cooperation between humans and technologies has to be taken into account to design work processes to developing productive synergies.

2.2 *Implementation*

Once new technologies are designed, the next step is to implement them in work processes [29]. In this context, the human factor has to be considered likewise. This concerns the information and communication about new technologies, to promote the workers' insights for example about the necessity and the objective of improving the work process as well as the workers' involvement and participation in the process.

This phase is about promoting the need for a change in work processes and behaviors to support the implementation of new technologies. In this context, the role of supervisors is crucial since they guide changes, answer questions, create space to try things out and gather ideas and suggestions of workers concerned. Change management elements are also possible instruments to be used in this phase [21].

2.2.1 Information and Communication

The success of implementing new technologies in work processes is highly determined and influenced by the available information for individuals [15]. Therefore, questions of information asymmetries, information policies and management are crucial for the perception of digital work [24, 46]. The work system framework as a socio-technical system plays an important role, as information is not only the basis for digital systems like using mobile handheld scanners for example. Information represents also the meta-level of information received about digital changes and projects, about expectations and objectives of technology implementation within supply chains. In this sense, communication as the exchange or transmission of information is highly relevant to design the change process, also the communication enabled with others – colleagues, experts and supervisors – about new technologies. The more structured and relevant information is available for workers and the better relevant communication is designed, the higher is the success probability of implementing new technologies.

2.2.2 Insights

Information and communication should promote the workers' understandings and create insights. Insights mean being able to comprehend the process change, the objective and purpose of introducing new technologies [24, 46]. In this respect, insights imply a similar way of assessing and evaluating a situation or necessity. It is not only about understanding the content of the innovation, but especially about being able to bring about a change in workers' actions and to improve their well-being and working life [30]. In this sense, workers need a deeper understanding of how and why new technologies are to be used and what the resulting effects are. Not least, this also supports workers' sense of control at work as the degree to which people have the feeling to possess control over outcomes of developments and processes [31].

2.2.3 Involvement

In the course of improving work processes, it is decisive to consider workers' perspectives [41]. Digitalization in organizations is usually a top-down process initiated by managers, but the acceptance of new technologies is higher if workers participate in the change process and their ideas are taken into account also bottom-up [56].

In cooperation with leaders and experts, work processes can jointly be selected for further improvement, digitalization options assessed and solutions for staff-oriented work design chosen. This can be connected to the discussion about prosuming as a combination of consumer and producer which has gained in importance especially in digital contexts [6, 48]. Moreover, the workers' participation promote the acceptance of new technologies, their self-efficacy, motivation and engagement [31].

2.3 Utilization

When new technologies are designed and implemented, they are used in daily business. The use of new technologies does not only depend on the design and implementation process but also on the usability of technologies and the integration into the overall work process. Usability consists of efficiency, effectiveness and satisfaction [36]. This will also affect the cooperation of workers with their colleagues and their supervisors. The careful integration into work processes promotes the workers cooperation and motivation to engage in the implementation of new technologies. As a result, the workers' acceptance or reluctance as well as their motivation becomes relevant and decisive for the success of implementing new technologies.

2.3.1 Acceptance and Reluctance

For the use of new technologies, the process of implementation has to consider certain aspects to promote the workers' acceptance instead of their reluctance. According to Rogers [49], they need knowledge about the change expected and persuasion through communication. To promote the adoption of new technologies, workers need to see the relative advantage of using them and they need the possibility to test the innovations. The Technology Acceptance Model [55] argues that technology use is directly determined by behavioral intention to use it and this depends of the *perceived usefulness* of the technology (i.e., the utility or instrumentality of the technology for goal attainment) and *perceived ease of use* (i.e., the usability of the technology). Not least, for the acceptance of new technologies also trust is needed [33]. This is important as trust has a multi-facetted positive impact on technology acceptance: First, trust is supporting acceptance of specific new technologies eo ipso. Second, trust can transfer successful experiences with new technologies in one case towards another new case of technology implementation. Therefore, trust-building and trust-oriented planning actions are important parts in digital implementation processes [12].

2.3.2 Motivation

Motivation is central to dealing with changes in work contexts. Also in the course of implementing new technologies, workers' motivations are affected [54]. As especially intrinsic motivation promotes high performance, certain measures are to be considered [17]. In this sense, the experienced meaningfulness, responsibility for outcomes of work as well as knowledge of actual results of work activities have to be maintained and provided through the use of new technologies. It remains unclear how the use of new technologies can promote motivation [54]. According to the Self-Determination Theory [51], a context that provides sufficient possibilities to experience autonomy will give rise to intrinsic motivation, whereas a context that is perceived as excessively controlling can impair motivation. Autonomy in the sense of experience of choice, personal agency, volition, self-determination, competence and relatedness is a crucial psychological need in these contexts. Especially the technology that addresses these needs will facilitate intrinsic motivation. It would be essential to align work tasks and personal goals as well as to provide workers with opportunities for an effective, autonomous performance [54]. In this sense, also empowerment refers to strategies and measures that are intended to increase the degree of autonomy and self-determination of workers and enable them to represent their interests autonomously, responsibly and self-determinedly [47, 53].

2.3.3 Cooperation

The use of new technologies in work contexts also affects workers' cooperation. In socio-technical systems, coordination is important for problem solving and with this the ability to work closely, democratically and creatively with others [41]. Thus, cooperation with colleagues changes, also because work tasks can be performed alone without the help of others. Second, also the work relationship with supervisors might change since interaction take place digitally as well as the recording of performance measures. Supervisors can monitor and assess the fulfillment of work tasks with the help of new technologies and supervise workers' performances [26]. Third, also cooperation of humans with new technologies (HCI) are of high interest and relevant to the success of the work process. We put the HCI as cooperation to underline that working with technologies also requires some form of interaction [44]. This implies a division of work tasks in the sense of who is in charge of making decisions based on the information provided.

2.4 Evaluation

In a final phase, the qualitative and quantitative impact of a change through the implementation of new technologies has to be assessed. This concerns workers' improvements especially regarding their competencies, self-efficacy, motivation, work behavior and satisfaction. Thus, to achieve a motivated workforce, the socio-technical system is expected to address goals and measures relevant in the workers' perspectives [30]. Self-efficacy is key for a confident attitude towards new technologies and the ability to cope with changing work processes [4]. Also, the (altered) performance is to be considered to evaluate the use of new technologies in a quantitative and qualitative manner.

2.4.1 Improvement

One key factor to estimate the success of implementing new technologies is the improvement of workers and work processes. The use of new technologies and change of work processes implies the opportunity to improve working conditions and the workers' well-being. However, it also requires different competence profiles of workers since human abstraction and problem-solving skills in dealing with complexity are necessary. The balance between altered work requirements on the one hand and required competencies on the other hand is critical for the workers' improvement [2]. The workers' competence development can also be evaluated by considering their self-efficacy. Especially the workers' self-efficacy is crucial for coping behavior and effort as it is concerned with individual beliefs about ones capability to cope with certain and new situations, to use ones abilities to solve problems and tasks, e.g. whether they feel competent in using and learning how to use new technologies [4, 27]. Self-efficacy can be fostered through performance accomplishments, vicarious experience, verbal persuasion and emotional arousal – all measures to promote dealing with new technologies in work processes.

2.4.2 Performance

Finally, to estimate the success of implementing new technologies, the performance of workers in digitalized work processes is to be analyzed. The use of socio-technical systems is regarded as a means of promoting to achieve organizational goals. We differentiate quantitative and qualitative measures. Quantitative measures imply higher quantities and greater efficiency of work processes; qualitative measures refer to a better production quality and the number of idea submissions. Productivity is especially enforced through the congruent pursuit of workers' and organizational goals [15]. To measure performance, workers' surveillance and monitoring processes come into play. Electronic performance monitoring systems have negative consequences on job satisfaction, commitment and self-efficacy [22]. Thus, automation and digitalization help to monitor workers' performance more closely [40]. Moreover, new technologies also promote algorithmic control [34, 38], meaning control from the system used and its algorithm. These individual performance metrics also support self-control of workers and the comparison of individual engagements to improve performance [9]. However, the respective data protection regulations must be taken into account. This is addressing the adherence to legal regulations and laws but also goes beyond this legal perspective as workers' perceptions and perspectives e.g. regarding being controlled through data acquisition have to be included in a comprehensive evaluation [3, 11].

3 Contribution Overview and Chapter Structure

This book compiles a sequence of chapters dealing with the outlined topic of digital supply chains in order to provide a structured synopsis to consider the human factor in analyzing digital work processes. This stems among others from several scientific events and networks as for example a Club of Florence research workshop held at the international: "University of Göttingen" in December 2018 as well as the 2020 VHB conference at the Goethe University Frankfurt in Germany.

Contributions incorporate a wide variety of disciplines as well as from academic and business practice perspectives. The basic idea for this setup is to include different aspects and central perspectives for considering the role of the human factor in digital work processes. The contributions are organized according to the structured synopsis as outlined in the following.

3.1 Design Phase

In the first part, we introduce two chapters addressing the design phase of new technologies: The contribution of Semhar Kinne, Jana Jost, André Terharen, Felix Feldmann, Michael Fiolka and Thomas Kirks deals with a top-down process development system for CPS design and integration in the fourth industrial revolution. They present an analysis of existing technologies, design methods and system description techniques that simplifies the design of logistics systems along processes, tasks and implementations under the premise of HCI. They apply their mode to intralogistics process of order picking and provide multiple variations of HCI technologies. The chapter by Hendrik Lager, Alfredo Virgillito and Tom-Philipp Buchberger deals with ergonomic improvements on the one hand and work intensification through digitalization in logistics on the other hand. They present qualitative-explorative results from different logistics departments and warehouses using digital technologies in order picking. They show that substitution potentials are almost exclusively used for intensifying work and hence, that possible positive ergonomic effects are eliminated or not at all a crucial aim when digital technologies are implemented in order picking. This highlights the challenges and demands envisioned in the structured synopsis in order to arrive at positive and productive human-centered digital work settings in supply chains.

3.2 Implementation Phase

In the second part, we cluster chapters on the implementation of new technologies in organizations: The contribution of *Martin Eisenmann* and *Nick Große* deals with digital twins in industrial maintenance. The growing data structure, various functionalities of digital technologies increase complexity in industrial maintenance. The authors reflect on the consequences of using these technologies for human labor. This is of utmost importance as the outlined elements of information, insight and involvement are addressed as well as their importance for productive supply chain processes. The chapter by *Ingo Börsting* and *Marc Hesenius* does discuss the use of chatbots in digital work environments and presents a systematic approach for chatbot development. They emphasize that chatbots typically have to be configured with knowledge about the use case at hand, potential dialog paths and user interactions, which means to explicitly define user input and chatbot response. In this context, also the interplay between humans and machines as well as the perceived user experience (UX) matters. Altogether, the two chapters for the implementation phase transport the message of human-centered approaches although very technical computer science aspects are likewise addressed.

3.3 Utilization Phase

The third part is about the use of new technologies with reference to the components as outlined in the structured synopsis. The contribution of *Caroline Ruiner* and *Christina Debbing* does focus on the adoption process of digital technologies and analyzes workers' attitudes towards new technologies in digital supply chains. The focus is on antecedents for a fruitful collaboration, aiming at the picture that they become friends instead of foes. They apply the Theory of Innovations [49] to the discussion on digitalization, present empirical findings regarding digitalization processes in an organization and identify antecedents for a gainful collaboration of people working with digital devices. The chapter by *Andrea Bottalico* on automation processes and industrial relations in the port industry reveals how a technological innovation introduced profoundly changes the logistics labor market. Thus, digitalization and automation processes affect working conditions, job profiles, training systems and skills required for the maritime-logistics chain. The contribution of *Mona Bardmann* deals with intercultural teamwork in multinational enterprises as this is of high relevance in globalized supply chains. She focuses on the cooperation amongst and in teams and the resulting formation of relationship networks, how team leaders and team members cooperate in these settings. Based on a qualitative study, she outlines a permeable unified storage culture which leads to a unification of differences in digitalized work contexts. Altogether, human involvement in the utilization is conceptualized as crucial for form and process performance, especially in digital settings in this section.

3.4 *Evaluation Phase*

The fourth and last phase is dedicated to the evaluation of implementing new technologies in organizations and how this relates to the human factor. The contribution of *Dominic Loske* and *Matthias Klumpp* reflects on efficiency measurement in digitalized work systems of transport logistics. They use the data envelopment analysis method to devise a quantitative efficiency calculation of the truck driver work system with specific input and output types. Their results support the notion of crucial human worker involvement in the changeover to new digital technology implementation in supply chain processes. The contribution of *Riccardo Emilio Chesta* analyzes the cycle of worker contention at Amazon logistics in Italy and the link between digitalization in logistics and worker mobilization. He puts emphasis on how technology has affected working conditions as well as how worker organizing triggered specific forms of unionism. His analysis reveals how technological innovation can be negotiated as part of collective bargaining where the regulation of an employment relations system plays a central role. Obviously, this section regarding evaluation is critical to long-term success of digital process changes.

With these different chapters, we provide insights into the relevance and the role of the human factor in digitalization processes in supply chain work environments. Altogether, we are in the process of learning that the focus on technical aspects alone is not enough. Rather the human workers, their expectations, needs, ideas and contributions have to be considered to establish successful human-machine-teams in work contexts. This could lead to win-win-situations for workers and organizations alike throughout global supply chains. With the changes in work processes and a profound change in the world of work, it will be increasingly important to offer attractive working conditions and design sustainable employment relationships. This book provides a first glance into the details of the human factor in digital supply chain processes and will continue an ongoing discussion regarding this crucial topic for successful digital transformations in organizations.

References

1. Autor DH (2015) Why are there still so many jobs? The history and future of workplace automation. J Econ Perspect 29(3):3–30
2. Bakker AB, Demerouti E (2007) The job demands-resources model: state of the art. J Managerial Psychol 22(3):309–328
3. Ball K (2010) Workplace surveillance: an overview. Labor History 51(1):87–106
4. Bandura A (1977) Self-efficacy: toward a unifying theory of behavioral change. Psychol Rev 84(2):191–215
5. Baxter G, Sommerville I (2011) Socio-technical systems: from design methods to systems engineering. Interact Comput 23(1):4–17
6. Beer D, Burrows R (2010) Consumption, prosumption and participatory web cultures: an introduction. J Consum Culture 10(1):3–12
7. Carroll JM (1997) Human–computer interaction: psychology as a science of design. Int J Hum Comput Stud 46(4):501–522

8. Cummings M, Bruni S (2009) Collaborative human-automation decision making. In: Nof S (ed) Handbook of automation. Springer, Berlin, pp 437–447
9. Dale S (2014) Gamification: making work fun, or making fun of work? Bus Inform Rev 31(2):82–90
10. Dalenogare LS, Benitez G, Ayal NF, German A (2018) The expected contribution of Industry 4.0 technologies for industrial performance. Int J Prod Econ 204(C):383–394
11. Day A, Paquet S, Scott N, Hambley L (2012) Perceived information and communication technology (ICT) demands on employee outcomes: the moderating effect of organizational ICT support. J Occup Health Psychol 17(4):473–491
12. De Filippi P, Mannan M, Reijers W (2020) Blockchain as a confidence machine: the problem of trust & challenges of governance. Technol Soc 101284
13. Dul J, Bruder R, Buckle P, Carayon P, Falzon P, Marras WS, van der Doelen B (2012) A strategy for human factors/ergonomics: developing the discipline and profession. Ergonomics 55(4):377–395
14. Edwards P, Ramirez P (2016) When should workers embrace or resist new technology? New Technol Work Employment 31(2):99–113
15. Elbert KK, Kroemer HB, Hoffman ADK (2018) Ergonomics: how to design for ease and efficiency. Academic, New York
16. Frey CB, Osborne MA (2013) The future of employment: how susceptible are jobs to computerisation. OMS Working Paper
17. Hackman JR, Oldham GR (1975) Development of the job diagnostic survey. J Appl Psychol 60(2):159–179
18. Hansen HK, Flyverbom M (2015) The politics of transparency and the calibration of knowledge in the digital age. Organization 22(6):872–889
19. Hartson R, Pyla PS (2012) The UX book: process and guidelines for ensuring a quality user experience. Elsevier, Amsterdam
20. Holford WD (2019) The future of human creative knowledge work within the digital economy. Futures 105:143–154
21. Jafari P, Mohamed E, Lee S, Abourizk S (2020) Social network analysis of change management processes for communication assessment. Autom Construct 118:103292
22. Jeske D, Santuzzi AM (2015) Monitoring what and how: psychological implications of electronic performance monitoring. New Technol Work Employment 30(1):62–78
23. Kirchner S, Schüßler E (2019) The organization of digital marketplaces: unmasking the role of internet platforms in the sharing economy. In: Ahrne G, Brunsson N (eds) Organization outside organizations: the abundance of partial organization in social life. Cambridge University Press, Cambridge, pp 131–154
24. Kitchen PJ, Daly F (2002) Internal communication during change management. Corp Commun Int J 7(1):46–53
25. Klumpp M (2018) Automation and artificial intelligence in business logistics systems: human reactions and collaboration requirements. Int J Logistics Res Appl 21(3):224–242
26. Klumpp M, Ruiner C (2018) Digitalization and work organization in new urban food delivery systems. Int J Food Syst Dyn 9(5):399–408
27. Klumpp M, Ruiner C (2019) Human role in digital logistics: relevance of intuition in interacting with AI. In: Bierwirth C, Kirschstein T, Sackmann D (eds) Logistics management. Lecture notes in logistics. Springer, Cham, pp 32–44
28. Klumpp M, Zijm H (2019) Logistics innovation and social sustainability: how to prevent an artificial divide in human-computer interaction. J Bus Logistics 40(3):265–278
29. Kong XTR, Zhong RY, Zhao Z, Shao S, Li M, Lin P, Chen Y, Wu W, Shen L, Yu Y, Huang GQ (2020) Cyber physical ecommerce logistics system: an implementation case in Hong Kong. Comput Ind Eng 139:106170
30. Land F (2000) Evaluation in a socio-technical context. In: Baskerville R, Stage J, DeGross JI (eds) Organizational and social perspectives on information technology. Springer, Boston, pp 115–126

31. Larjovuori RL, Bordi L, Mäkiniemi JP, Heikkilä-Tammi K (2016) The role of leadership and employee well-being in organizational digitalization. Paper presented at the 26th Annual RESER Conference, Naples
32. Lee J, Seppelt B (2009) Human factors in automation design. In: Nof S (ed) Handbook of automation. Springer, Berlin, pp 417–436
33. Lee JD, Lee KA (2004) Trust in automation: designing for appropriate reliance. Human Factors J Human Factors Ergon Soc 46(1):50–80
34. Lee MK, Kusbit D, Metsky E, Dabbish L (2015) Working with machines: the impact of algorithmic and data-driven management on human workers. Paper presented at the ACM, Soul
35. Legenvre H, Henke M, Herbert R (2020) Making sense of the impact of the internet of things on purchasing and supply management: a tension perspective. J Purchas Supply Manage 26(1):100596
36. Lindgaard G, Dudek C (2003) What is this evasive beast we call user satisfaction? Interact Comput 15(3):429–452
37. Manavalan E, Jayakrishna K (2019) A review of internet of things (IoT) embedded sustainable supply chain for industry 4.0 requirements. Comput Ind Eng 127:925–953
38. Mohlmann M, Zalmanson L (2017) Hands on the wheel: Navigating algorithmic management and Uber drivers' autonomy. Paper presented at the proceedings of the international conference on information systems (ICIS 2017), Seoul, South Korea
39. Montreuil B (2011) Towards a physical internet: meeting the global logistics sustainability grand challenge. Logistics Res 3(2):71–87
40. Moore PV, Upchurch M, Whittaker X (2018) Humans and machines at work: monitoring, surveillance and automation in contemporary capitalism. In: Moore PV, Upchurch M, Whittaker X (eds) Humans and machines at work: monitoring, surveillance and automation in contemporary capitalism. Springer International Publishing, Cham, pp 1–16
41. Mumford E (2000) A socio-technical approach to systems design. Requirements Eng 5(2):125–133
42. Norman DA, Stappers PJ (2015) DesignX: complex sociotechnical systems. She Ji J Design Econ Innov 1(2):83–106
43. Orlikowski W (2016) Digital work: a research agenda. In: Czarniawska B (ed) A research agenda for management and organization studies. Edward Elgar Publishing, Northampton, pp 88–96
44. Orlikowski WJ (1992) The duality of technology: rethinking the concept of technology in organizations. Organ Sci 3(3):398–427
45. Phan S, Ballot E, Huang GQ, Montreuil B (2017) Physical internet and interconnected logistics services: research and applications. Int J Prod Res 55(9):2603–2609
46. Proctor T, Doukakis I (2003) Change management: the role of internal communication and employee development. Corp Commun Int J 8(4):268–277
47. Rappaport J (1981) In praise of paradox: a social policy of empowerment over prevention. Am J Community Psychol 9(1):1–25
48. Ritzer G, Jurgenson N (2010) Production, consumption, prosumption: the nature of capitalism in the age of the digital 'prosumer.' J Consumer Culture 10(1):13–36
49. Rogers EM (2003) Diffusion of innovations. Free Press, New York
50. Rousseau DM (1995) Psychological contracts in organizations: understanding written and unwritten agreements. Sage, Thousand Oaks
51. Ryan RM, Deci EL (2000) Self-determination theory and the facilitation of intrinsic motivation, social development and well-being. Am Psychol 55(1):68–78
52. Sin J, Munteanu C (2020) An empirically grounded sociotechnical perspective on designing virtual agents for older adults. Human–Comput Interact 35(5–6):481–510
53. Spreitzer GM (1995) Psychological empowerment in the workplace: dimensions, measurement, and validation. Acad Manage J 38(5):1442–1465
54. Szalma JL (2014) On the application of motivation theory to human factors/ergonomics: motivational design principles for human–technology interaction. Human Factors J Human Factors Ergon Soc 56(8):1453–1471

55. Venkatesh V, Bala H (2008) Technology acceptance model 3 and a research agenda on interventions. Decis Sci 39(2):273–315
56. Vink P, Imada AS, Zink KJ (2008) Defining stakeholder involvement in participatory design processes. Appl Ergon 39(4):519–526
57. Wickens CD, Gordon SE, Liu Y, Lee J (2004) An introduction to human factors engineering. Prentice Hall, Upper Saddle River
58. Wilkesmann M, Wilkesmann U (2018) Industry 4.0—organizing routines or innovations? VINE J Inform Knowl Manage Syst 48:238–254
59. Wood AJ, Graham M, Lehdonvirta V, Hjorth I (2018) Good gig, bad gig: autonomy and algorithmic control in the global gig economy. Work Employ Soc 33(1):56–75
60. Yadav G, Luthra S, Jakhar SK, Mangla SK, Rai DP (2020) A framework to overcome sustainable supply chain challenges through solution measures of industry 4.0 and circular economy: an automotive case. J Cleaner Prod 254. https://doi.org/10.1177/0972150920948818
61. Zuboff S (1988) In the age of the smart machine: the future of work and power. Basic Books, New York

Matthias Klumpp is working at the Georg-August-University of Göttingen, FOM University of Applied Sciences Essen and Fraunhofer Institute for Material Flow and Logistics (IML) Dortmund. He received his Ph.D. from the University of Leipzig, his habilitation from the University of Duisburg-Essen and was in visiting positions at University of Twente, EUI Florence, University of Lugano (IDSIA), and University of Modena. Addressed research topics are sustainable, digital (AI), and resilient supply chains with a focus on HCI and qualification.

Caroline Ruiner is professor for sociology at the University of Hohenheim in Germany. She studied sociology and management science at Goethe-University Frankfurt, received her doctorate in sociology at the University of Augsburg and habilitated at the Faculty of Social Sciences at Ruhr University Bochum. Her research focuses on new work and new employment relationships and the effects on individual and organizational levels as well as with regard to industrial relations.

Design

Process Development for CPS Design and Integration in I4.0 Systems with Humans

Semhar Kinne, Jana Jost, André Terharen, Felix Feldmann, Michael Fiolka, and Thomas Kirks

Abstract This contribution presents an analysis of existing technologies, state-of-the-art design methods and system description techniques for a novel method that simplifies the design of logistic systems along processes, tasks and implementations under the premise of Human-Technology Interaction. For that objective, various characteristics of Human-Technology Interaction devices such as handhelds or smartphones and their interactions to define a Human-Technology Interaction classification are investigated. Furthermore, this approach connects the Distribution Center Reference Model to a common process description especially for distribution centers using the Business Process Model and Notation. A proposed interconnection between reference model, notation and classification allows a tangible cause and effect view on the deployment of devices with different characteristics of interaction. The findings are applied to the intralogistics process of order picking and provide multiple variations of Human-Technology interaction technologies. The resulting complex system requires an expert system, which can derive requirements to create a support system for all stakeholders.

Keywords Human-Technology Interaction · Warehouse logistics · Digitalization · Business process modelling

1 Introduction

New and innovative technologies in the context of Internet of Things and the fourth industrial revolution (Industry 4.0) enable typical logistic key figures such as reliability of delivery or service level to be optimized [11]. Despite an intended high level of automation, human employees with their versatile cognitive and fine motor skills

S. Kinne · J. Jost · M. Fiolka · T. Kirks (✉)
Fraunhofer Institute for Material Flow and Logistics (IML), Dortmund, Germany
e-mail: thomas.kirks@iml.fraunhofer.de

A. Terharen · F. Feldmann
TU Dortmund University, Dortmund, Germany

M. Klumpp and C. Ruiner (eds.), *Digital Supply Chains and the Human Factor*, Lecture Notes in Logistics,
https://doi.org/10.1007/978-3-030-58430-6_2

will continue to be an essential component of Cyber-Physical Systems (CPS) for the time to come. Technology systems therefore must dynamically adapt to these human actors in order to make optimal use of the strengths of both sides. Formerly, purely technological systems are being transformed into a socio-technical system consisting of the elements human, technology and organization. These elements cannot be considered independently of each other and require new ways of communication and interaction between humans and machines [10]. Due to cost pressure and performance requirements, the logistic industry is particularly challenged to provide efficient solutions. In operational business, it is almost impossible for companies to foresee the effects of the use of new Human-Technology Interaction (HTI) devices (e.g. smart watches, augmented reality (AR) and virtual reality (VR) technologies) in their complex process flows and to integrate them adequately. So far, there is no expert system that links existing models for process description with the characteristics of the selected HTI technology and shows process changes in relation to target variables such as lead time or quality. Such an expert system would provide guidance for decision-making towards using HTI devices. A holistic classification approach is required as prerequisite to match the characteristics of a certain device with a process model. The holistic classification approach is evaluated using a typical logistic process.

2 State of the Art

The following part gives an introduction to the field of human-technology interaction, which is necessary for the understanding of the introduced classification approach.

2.1 Human-Technology Interaction

There is a variety of definitions for the interaction between humans and technology. Often the terms "communication" and "interaction" are used synonymously as well as the wordings "Human-Computer Interaction", "Human-Machine Interaction" and "Human-Robot Interaction" [1, 8, 10]. To have a common understanding of the interaction process, the following definition of the center of excellence logistics and IT [10] is being used in this article. "Human-Technology Interaction", the superordinate of the previous terms, contains the interaction of all technological devices which are able to correlate their actions depending on their interaction partner. The basic requirement for interaction is the ability to communicate, i.e. to send and receive messages via at least one information channel (visual, acoustic, etc.) and includes additionally reacting on, dealing with and influencing one another, possibly even controlling each other and adapting to one another.

To characterize the interaction process and identify requirements which the suitable technology should fulfil, various classification approaches for HTI with different focuses have been proposed. Jost et al. [10] classify the interaction process according to the design form into four classes: information exchange, reaction, control and adaptivity. Onnasch et al. [13] propose an approach which classifies human-robot interaction regarding different categories e.g. form of interaction, role during interaction and communication channels [13]. Other categories in classification approaches can be found with Rammert [15]. The author divides HTI in the dimension of activity, characteristics of autonomous machines, level of ability and relationship between human and technology. Smithers [17] as well as Franklin and Graesser [6] also classify technology according to its autonomy (types of autonomous mechanism and characteristics of autonomous agents).

3 Holistic Classification Approach of HTI

The following section shows the way from applying the taxonomy to a use case to the required complements of a process view and an area of application specific classification model. This leads to the Holistic Classification Approach (HCA) of HTI, which is exemplarily applied to a complex intralogistics use case.

3.1 Taxonomy for HTI

Due to the large amount of technologies available at the market and therewith the rising number of classification approaches for HTI, the aim was to develop a generalized classification approach. This approach should be based on common literature [3, 10, 14, 15, 20] and likewise should not be limited to one kind of technology e.g. robots or automated guided transport vehicles. Further, the approach should offer the possibility to locate future technologies for HTI.

The HTI classification approach presented below is on the one hand intended to help technology developers to classify their technologies. On the other hand, future users of technology are able to find a suitable technology using the previously explained process view and the approach presented here. Further, it should be noted that a technology can be assigned to different classes depending on its area of application and its functionality. Therefore, it can be located at different positions in the morphological box.

Figure 1 shows the HTI classification approach applied to the cellular transport vehicle (CTV) which consists of different approaches each depicted row-wise. The designation of the approach can be found in the first column of the corresponding row, followed by the different characteristics in the following columns. Besides, there are short descriptions and complements in each cell. Furthermore, correlations between classes-if possible-are listed. There are classes which build on one another or can be seen as a further stage of the other and thus include the characteristics of

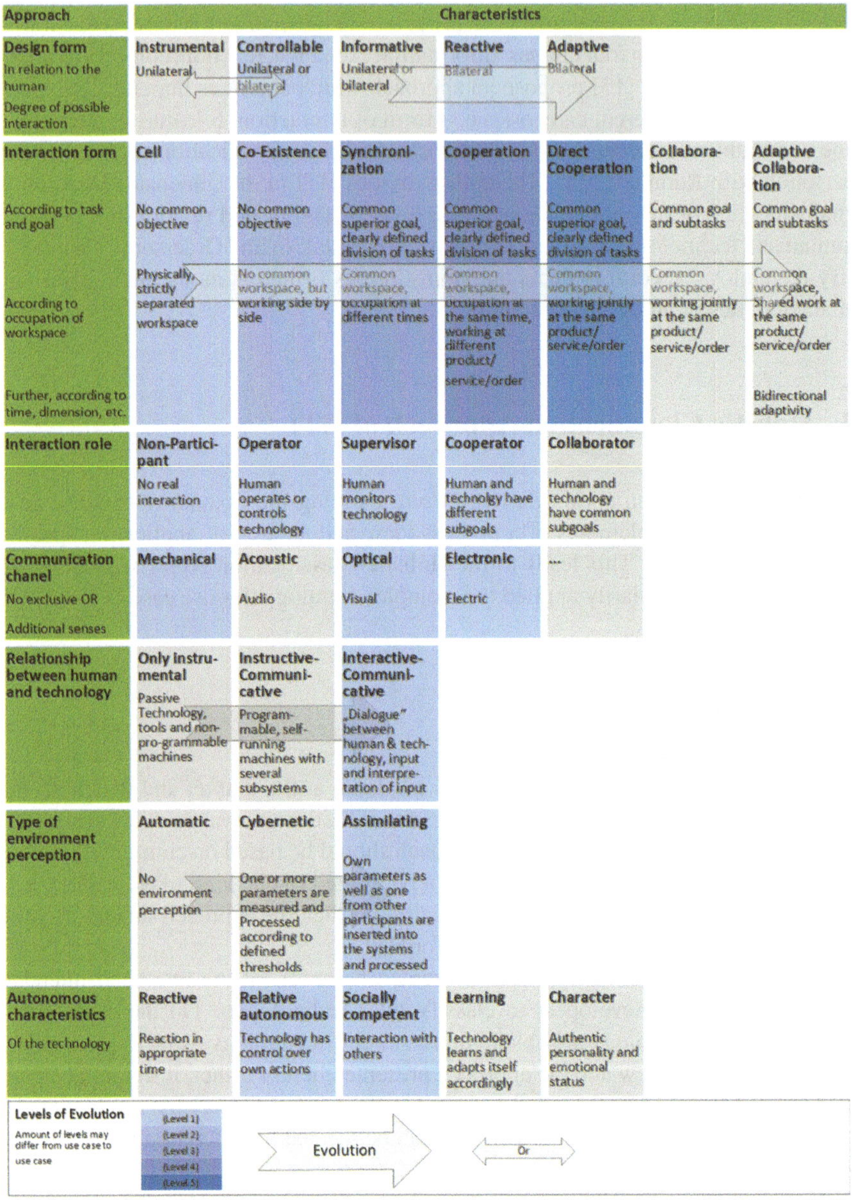

Fig. 1 Classification approach for HTI applied to cellular transport vehicle

the previous class (evolution level; these are given by different color levels of blue-these can be fewer or more, depending on the intended count of the users' levels of evolution). Others may be mutual exclusive. In addition to the classification degrees of the respective classification approach found in the literature, further classification degrees are being acquired.

The classification approach is under ongoing investigation and is merged from different research based approaches. The goal is to find a holistic classification model for technologies in HTI. The row design form is based on Jost et al. [10] four forms of design, which describes controllable, informative, active and reactive design structures. In addition, the form instrumental was added to also include passive interaction modalities with tools like screw drivers or drill hammers without outgoing information or sensory data processing. The row forms of interaction can be investigated under two perspectives, which are combined in a single row. Based on task and goal of interactions, the forms collaboration, cooperation and co-existence were proposed by Onnasch et al. [13]. Regarding the occupation of the workspace, Bender et al. [2] described the interaction forms cell, co-existence, synchronization cooperation and collaboration. Then, direct cooperation was added to describe a cooperative task on the same product. Furthermore, adaptive collaboration was added to the model, to differentiate between an adaptive system and a non-adaptive system. Onnasch et al. [13] also defined the roles of interaction for humans, which are described by the third row. They divided the roles into non-participant, operator, supervisor, cooperator and collaborator. Regarding the channel of communication, they derive the channels mechanical, acoustic, optical and electronic. More channels are possible, like olfactory or electro-encephalographic means of communication. As for the relationship between humans and technology, Rammert [15] delivered three different types of relationships. This can be purely instrumental with passive instruments involved, instructive-communicative through programmable machines or interactive-communicative, which requires a kind of dialog between actors. The latter includes the possibility of a dissonance between input and expected output, information may be misunderstood in the interpretation process [15]. From the perspective of machines, the perception of the environment is influential on the interaction, as well. Smithers [17] differentiates between automatic and cybernetic technologies. An automatic system represents a system without reactions to influences, while cybernetic systems involve the processing of one or more parameters. This definition requires an extension for systems, which not only include foreign parameters in their behavior, but also change the actions of other actors by bringing in their own parameters [17]. The term assimilating, for describing the adaptiveness of multiple actors, is proposed for this matter. An example would be a system with autonomous swarm behavior and decentralized control mechanics. Franklin and Grasser [6] proposed several properties for autonomous agents, which are also included in the HTI classification model. Autonomous attributes, like reactiveness of systems, are important for describing technology. Reactive systems provide responses in a timely manner. The original definition of the term autonomous is very fuzzy, more suitable for the HTI model is the term relatively autonomous, proposed by Rammert [15], which grants the technology the control of their own actions [15]. Socially able describes the ability to interact and communicate with other actors (artificial or human) and is based on the definition by Jennings and Wooldridge [20]. The attribute learning is based on the approach presented by Weiß [18], which also includes the ability of adaption of machines [18]. Another trait of a machine may be a credible personality, which also represents an emotional state [6].

The HTI classification approach presented here may have to be updated over time as new forms of technology or interaction concepts may arise and might lead to new levels and approaches. The classification of a CTV was carried out as an example and can be found in Fig. 1 (marked in blue). It should be noted that the classification of technology might depend on the area of application. There are classification approaches occurring in one classification degree regardless of the application and others in which the application is decisive for the classification degree.

For example, CTV always have a reactive behavior and evade or stop irrespectively of the application and the role of the human worker as well as the interaction form. In contrast to that, the vehicle can be perceived as co-existing for a visitor of the facility, but cooperates with a picker. Therefore, these specific evolution levels, in which the technology has to be viewed differently depending on the application, are shown in a color shading.

4 The Necessity of a Process View

As described above, a crucial part of a classification system is that a technology cannot be classified without its application area, respectively its purpose. There are many different technologies which are classified differently in the taxonomy depending on their application, therefore a mapping of the taxonomy with a process and a corresponding classification system is needed. In the work of Wu et al. [21] a layered framework is proposed to discompose elements in a socio-technical system, especially for emergent behavior results. In this case parts of the framework exist of an information flow model and a classification model. For the HTI classification approach related aspects are needed. In this case a process view represents not only the process steps, but also the information flow, supplemented with an area of application specific classification system.

In the field of intralogistics (corresponding to the use case above with the CTV), a variety of languages and methods exist for the presentation, description and documentation of business processes. Most common are the Unified Modeling Language, the Event-driven Process Chain and the Business Process Model and Notation, which are briefly described below. These methods can be used to map processes in complex intralogistic systems as process sequences [16]. The Distribution Center Reference Model is a specialized classification system for warehouse located processes, therefore intralogistics.

4.1 Unified Modeling Language

The Unified Modeling Language (UML) was merged in 1997 based on three object-oriented modeling methods and since then further developed as an object-oriented modeling standard by the Object Management Group (OMG). UML offers 14 different diagram types for visualizing the object-oriented relations between different elements. The Activity Diagram is particularly suitable for the representation of processes in intralogistic systems, since it can be used to represent the behavior of the system. The activity diagrams consist of the elements' actions, activity nodes, flows, object nodes and control nodes [4].

4.2 Event-Driven Process Chain

The Event-driven Process Chain (EPC) was developed by the Institute of Information Systems at Saarland University in 1992 in cooperation with SAP AG for documenting business processes. The extended Event-driven Process Chain (eEPC) supplements the processes with additional elements such as organizational units, responsibilities or information/material/documents. The EPC represents a process as a sequence of events, functions, linking operators and process paths. The sequence of the individual process steps is illustrated by arrows and connectors that show for example decision paths. The individual functions can be described in detail by further elements, such as information input and output or the information systems and organizational units that are involved [7, 12].

4.3 Business Process Model and Notation

The Business Process Model and Notation (BPMN) is an information visualization technique for the specification of business processes. The notation uses the flowcharting approach, which is also used in the UML activity diagrams. However, it focuses on the business process related aspects. It provides a standard for all business stakeholders to define and describe processes on an equal level. BPMN was originally developed by the Business Process Management Initiative (BPMI) and is now maintained by the OMG. The latest version 2.0.2 was published in 2014, with minor changes to the 2.0.1 version, which is also ISO certified [9].

BPMN has four different groups of graphical elements for representing the business process. Flow objects (events, activities and gateways) are the main elements of the processes. They describe that something happens (events, e.g. start or end event) and that something is done (activities, e.g. tasks). Gateways represent the forking and merging of paths. Connections show the path between the flow objects, as sequence flow, communication or association. The swim lanes group the flow objects either in pools, which represent the different participants of the process, or in lanes, which are part of a pool and organize e.g. the roles or functions of a participant. Artifacts represent additional information like a data object [5].

Based on the possibility to outline standard intralogistic processes in a simple manner and easily integrate additional information in the processes, we decided to choose the BPMN for defining our processes.

4.4 Distribution Center Reference Model

The Distribution Center Reference Model (DCRM) provides a classification framework for storage and picking processes in a distribution center [19]. It is a task-oriented model and presents a standardized approach to compare, analyze and evaluate processes in intralogistic systems. For the benchmarking of storage and picking processes several different calculation models are used. These models contain a detailed layout description and the material and information flow of the systems. The DCRM is based on other analytical methods like Supply Chain Operations Reference Model (SCOR) and has a hierarchical composition with four levels (top, process, task and implementation) [19]. In our case the process level is relevant. The DCRM divides the process level in six sub-processes which are generally found in intralogistic systems.

The sub-processes are the following:

- Receiving
- Storage and picking
- Consolidation and packing
- Shipping
- Added value
- Overhead.

The first five sub-processes describe the material flow within intralogistic systems, whereas the overhead process is on a superordinate level and describes all administrative tasks. This publication focuses on the two main processes storage and picking, which are consolidated in the DCRM. Figure 2 shows the flow chart of the main

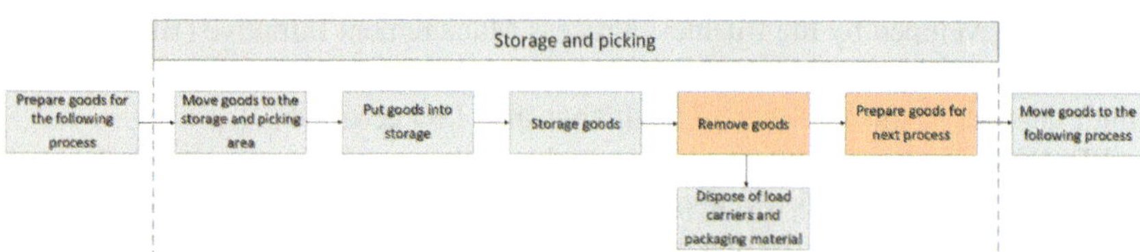

Fig. 2 DCRM flow chart of storage and picking process

activities within the consolidated process. The process steps analyzed later on are marked in orange.

Especially, the order picking process combines the usage of new technologies and the versatile cognitive and fine motor skills of the human workforce. That is why we chose to investigate the HTI on this process, described in the following.

5 Application of HCA in the Context of DCRM

In order to show the process changes resulting from using a specific HTI device, DCRM and BPMN methods are combined with the HTI classification approach. The individual process steps are shown in swim lanes (BPMN). The procedure is evaluated on the basis of an exemplary order picking process according to the person-to-goods principle. The application is first carried out for the process with a typical device for mobile data collection and then with the aid of Augmented Reality (AR) glasses. These devices are worn like glasses and allow information to be presented directly in the user's field of view. Depending on the specification and implemented functionalities of the glasses, different degrees of support through the technology are possible.

The process is categorized using the DCRM method as a storage and picking process with the characteristics "SP6: Storage and removal of single items" and "SP_B: Person to goods with one- or two-dimensional movement: Shelf storage". For the comparison between conventional and AR equipment, the sub-processes "remove goods" and "prepare goods for next process" are considered in particular. Figure 3 shows the order picking process by using a portable data terminal (PDT).

A picking order is initiated by PDT and requested by the worker. A picking list appears on the screen, as well as necessary picking aids that are required as load carriers or for transportation. The employee moves to the assigned location in the warehouse. This information is shared by specifying shelf, level and compartment. By scanning the labels, it is verified whether it is the correct article and if so, the withdrawal quantity is shown in the display. After the article has been removed and placed on the picking aid, the order item is confirmed. In this way, the entire picking order is compiled and finally made available at the collection point. The confirmation of arrival at the target position marks the end of the picking order process.

In the figure, actions that can be omitted when working with AR glasses are highlighted red, whereas green colored fields can be improved in time or quality.

The characteristics of PDT are shown in Fig. 4. PDT is a controllable device that transmits information to humans. The picking order and the workspace are the same for both interaction partners and there is division of tasks (human walks, picks and carries; PDT provides information and scans). For this reason, the working relationship is classified as direct interaction, in which the person operates the device. The PDT can communicate with the operator on different channels, requiring no haptic processes. The system works automatically and purely reactively based on a pre-programmed control.

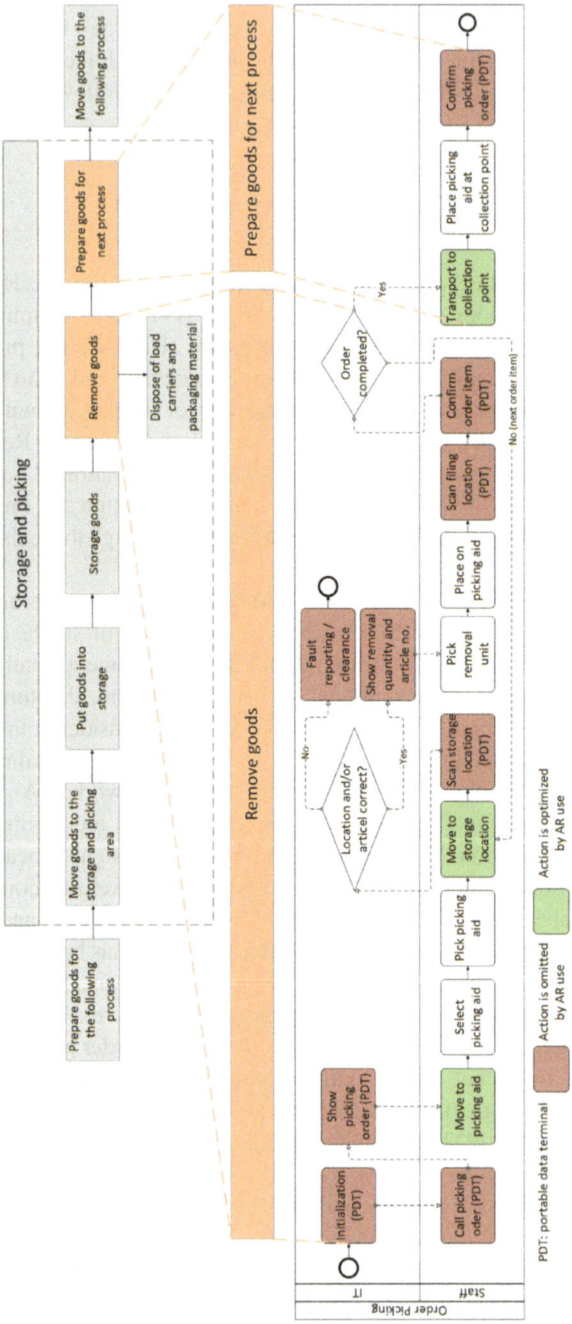

Fig. 3 DCRM and BPMN representation of an order picking process using a portable data terminal

Fig. 4 HTI classification of PDT in the exemplary order picking process

The properties of the process with AR glasses do not change with respect to the DCRM representation, but the category AR is added in the swim lane representation (Fig. 5). The AR glasses fulfill four different functions (navigation, confirmation, identification and highlighting), which in turn can have three levels of configuration. Both the function and the level of configuration influence the classification.

Figure 6 represents the HTI classification for AR glasses using the function navigation as an example. The glasses are a controllable device worn by the operator. The degree of possible interaction differs depending on the level of configuration. Level

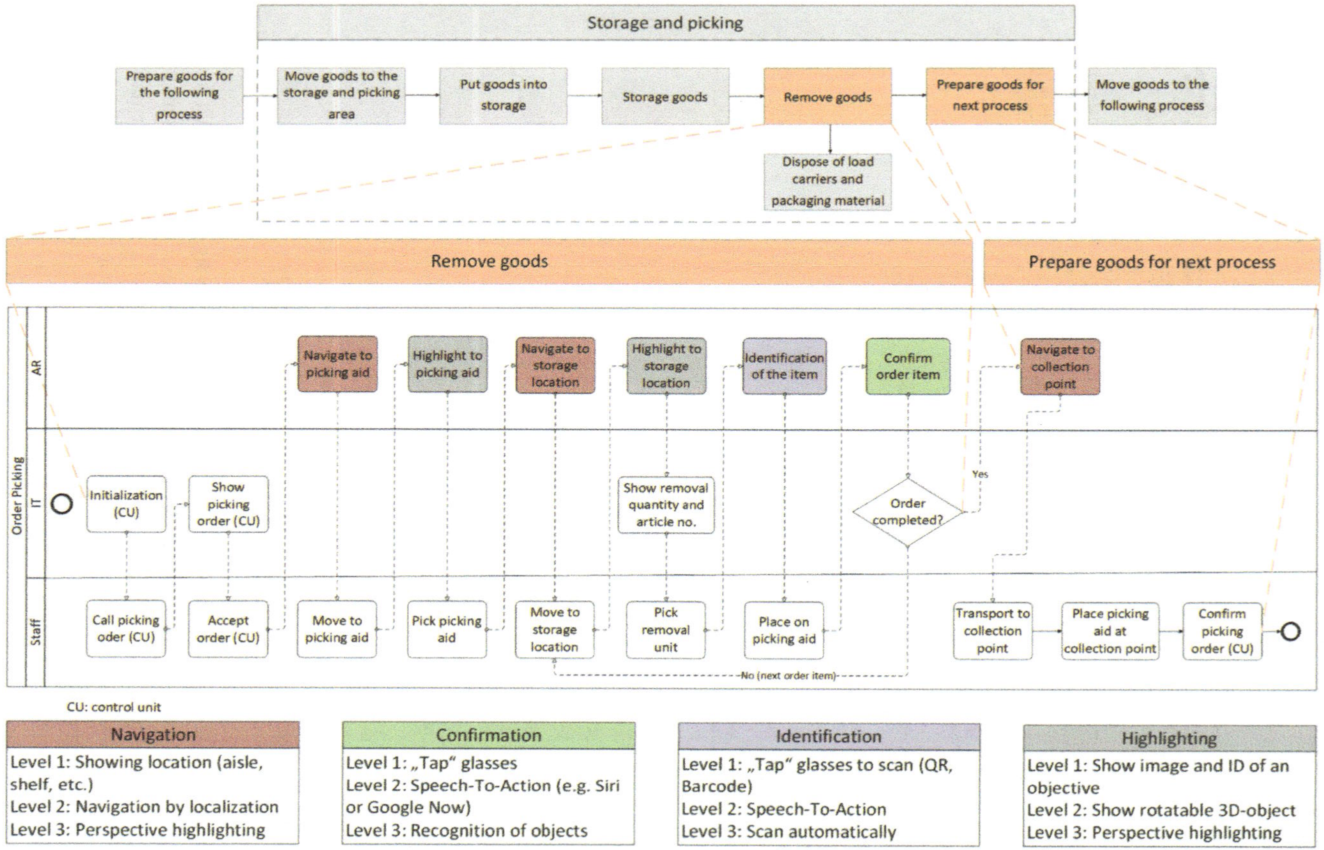

Fig. 5 DCRM and BPMN representation of an order picking process using AR glasses

Fig. 6 HTI classification of AR in the exemplary order picking process

1 is a pure presentation of information comparable with the use of PDT, whereas the view in Levels 2 and 3 adapts to the user's position reactively. As with the use of PDT, the form of interaction with AR glasses is a direct interaction. The tasks of the glasses and the human are clearly separated, but both act according to the superior picking task in the same workspace. Likewise, the classifications of interaction role, communication channel and the human-technology relationship are equally classified in comparison to PDT. In terms of the type of environmental perception, however, glasses operate at a higher evolutionary level and must be classified as cybernetic, since the current location is always important for navigation. A measurement of parameters is therefore assumed. The autonomous characteristics of the device differ according to the level. In Levels 2 and 3, the technology changes the arrangement of displayed information independently.

6 Conclusion and Outlook

The proposed HCA characterizes interaction processes and helps to identify requirements which a technology can fulfil. Various classification categories such as form of interaction, role during interaction and communication channels are taken into account to classify human-technology interaction. After performing a HTI classification, these technologies are matched to business processes in different levels of detail. This contribution provided examples that demonstrate a HCA application from beginning to the end. This approach helps multiple types of stakeholders. Technology developers are supported to classify their technologies in a structured manner. Process developers are given a tool to change processes in the course of digitalization more easily and deploy new technologies while recognizing different effects that are involved in the change. Managers can simulate changes before actually invoking them and they can see KPI change while altering the deployment of several technologies.

The depicted notation, classification and interconnection of technologies in the context of HTI is a complex method and demands the user to memorize a diverse amount of definitions and complex relations to transform default business processes to advanced, technology-enriched processes of better performance. This work may be time-consuming and challenging at the same time. To overcome this demanding issue, in future work one can derive requirements for a decision support system. This system can be fed with all concerning data and it will directly show resulting performance indicators like process times. Furthermore, future work can evaluate iterations of the developed system in near-industry setups with companies to obtain feedback and have it considered for further developments of the system. The DCRM is a functional process classification especially for distribution centers. Further work will also consider the portability of the relations between processes and technologies to gain a more generalized classification approach.

References

1. Bendel O (2017) Definition: Mensch-Maschine-Interaktion. Gabler Wirtschaftslexikon, Jan 26, 2017. [Online]. Available https://wirtschaftslexikon.gabler.de/definition/mensch-maschine-interaktion-54079/version-188399. Accessed 24 Feb 2020
2. Bender M, Braun M, Rally P, Scholtz O (2016) Leichtbauroboter in der manuellen Montage - einfach EINFACH anfangen
3. Bradshaw JM, Hoffman RR, Johnson M, Woods DD (2013) The seven deadly myths of autonomous systems. IEEE Intell Syst 28(3):54–61
4. Czuchra W (2010) UML in logistischen Prozessen: graphische Sprache zur Modellierung der Systeme; mit 4 Tabellen; [mit Online-Service]. 1. Ed. Vieweg + Teubner, Wiesbaden
5. Fischer L (ed) (2012) BPMN 2.0 handbook second edition: methods, concepts, case studies and standards in business process management notation. Future Strategies, Lighthouse Point
6. Franklin S, Graesser A (1997) Is it an agent or just a program? A taxonomy for autonomous agents. In: Müller JP, Wooldridge MJ, Jennings NR (eds) Intelligent agents III agent theories, architectures, and languages, vol 1193. Springer, Berlin, pp 21–35
7. Gadatsch A (2015) Geschäftsprozesse analysieren und optimieren. Springer Fachmedien Wiesbaden, Wiesbaden
8. Holzinger A (2001) Mensch-Computer-Interaktion (HCI) im 21. Jahrhundert und die Informatik in Österreich. e & i Elektrotechnik und Informationstechnik 118(3):147–149
9. ISO/IEC 19510:2013. Information technology—object management group business process model and notation. International Organization for Standardization, Genève ISO IEC 2013
10. Jost J, Kirks T, ten Hompel M. (eds.), Henke, M. (eds.), Clausen, U. (eds.) (2017) Herausforderungen der Mensch-Technik-Interaktion in der Intralogistik 3
11. Lieberoth-Leden C, Röschinger M, Lechner J, Günthner WA (2017) Logistik 4.0. In: Reinhart G (eds.) Handbuch Industrie 4.0: Geschäftsmodelle, Prozesse, Technik. Hanser, München, pp 451–606
12. Nüttgens M, Rump F (2002) Syntax und Semantik Ereignisgesteuerter Prozessketten (EPK), Bd. 21, pp 64–77
13. Onnasch L, Maier X, Jürgensohn T (2016) Mensch-Roboter-Interaktion - Eine Taxonomie für alle Anwendungsfälle. Bundesanstalt für Arbeitsschutz und Arbeitsmedizin (BAuA)
14. Parasuraman R, Sheridan TB, Wickens CD (2000) A model for types and levels of human interaction with automation. IEEE Trans Syst Man Cybern Part A Syst Humans 30(3):286–297
15. Rammert W (2003) Technik in Aktion: verteiltes Handeln in soziotechnischen Konstellationen (TUTS—Working Papers, 2-2003). Technische Universität Berlin, Berlin, p 24
16. Scholz-Reiter B, de Beer C, Freitag M, Hamann T (2008) Dynamik logistischer Systeme. In: Nyhuis P (ed) Beiträge zu einer Theorie der Logistik. Springer, Berlin, pp 109–138
17. Smithers T (1997) Autonomy in robots and other agents. Brain Cogn 34(1), 88–106
18. Weiß G (1996) Adaptation and learning in multi-agent systems: some remarks and a bibliography. In: Weiß G, Sen S (eds.) Adaption and learning in multi-agent systems. Springer, Berlin, pp 1–21
19. Wisser J (2009) Der Prozess Lagern und Kommissionieren im Rahmen des Distribution Center Reference Model (DCRM). Universitätsverlag Karlsruhe, Karlsruhe
20. Wooldridge M, Jennings NR (1995) Intelligent agents: theory and practice. Knowl Eng Rev 10(2):115
21. Wu PP-Y, Fookes C, Pitchforth J, Mengersen K (2015) A framework for model integration and holistic modelling of socio-technical systems. Decis Support Syst 71:14–27

Semhar Kinne studied mechanical engineering at TU Dortmund University, where she graduated in 2010. Since 2011, she is working as a research assistant at the Fraunhofer Institute for Material Flow and Logistics in the Department Machines and Facilities. She focuses on handling processes in intralogistics and researches different types of Human-Technology Interaction as well as the use of exoskeletons to improve physical ergonomics.

Jana Jost is a senior researcher and deputy head of the department Automation and Embedded Systems at the Fraunhofer Institute for Material Flow and Logistics (IML) in Dortmund. She received her master's degree in applied computer science majoring in robotics from the TU Dortmund University. Her research addresses human-technology interaction and ways to integrate the human in Industry 4.0 processes while focusing on mental and physical workload.

André Terharen is research assistant at the Chair of Materials Handling and Warehousing (FLW), Faculty of Mechanical Engineering, at TU Dortmund University. He researches in the field of virtual reality in intralogistics and is Scientific Coordinator of the Center of Excellence for Logistics and IT.

Felix Feldmann is a Ph.D. student at the Chair of Materials Handling and Warehousing at TU Dortmund University. He studied logistics at the TU Dortmund University and received his master degree there. He participates in the research project Centre of Excellence Logistics and IT and is part of the research clan Human-Machine-Interaction.

Michael Fiolka is a research assistant at the Fraunhofer Institute for Material Flow and Logistics Dortmund. He studied logistics and supply chain management at the TU Dortmund University and has worked as a student assistant at the Fraunhofer IML since 2017.

Thomas Kirks is a research assistant at the Fraunhofer Institute for Material Flow and Logistics at the department of Automation and Embedded Systems. He studied information technology at the Dortmund University of Applied Sciences and Arts and received his master's degree. Since 2012, he works as research assistance at the Fraunhofer Institute for Material Flow and Logistics in the field of Human-Technology Interaction and since 2017, he is the supervisor of the research clan Human-Technology Interaction at the Center of Excellence for Logistics and IT in Dortmund, Germany.

Digitalization of Logistics Work: Ergonomic Improvements Versus Work Intensification

Hendrik Lager, Alfredo Virgillito, and Tom-Philipp Buchberger

Abstract Digital technologies are more and more finding their way into logistics. On the one hand, they are associated with positive effects on the ergonomics of logistics work by replacing repetitive, stressful and ergonomically disadvantageous tasks. But, on the other hand, studies indicate that the intensification of work is increasing. According to this discrepancy, we investigate how digital technologies, ergonomic improvements and work intensification are related. We present qualitative-explorative results from different logistics departments and warehouses using digital technologies in order picking. Our aim is to analyze whether opportunities to substitute ergonomically disadvantageous tasks are used to integrate more challenging and attractive activities or whether they give rise to easier tasks. We show that such substitution potentials are almost exclusively used for intensifying work and hence, that possible positive ergonomic effects are eliminated or not at all a crucial aim when digital technologies are implemented in order picking.

1 Introduction

Digitalization and its technologies and solutions are often seen as the answer to cope with the increasing demands for individual products and services, shorter delivery times within dynamic environments and markets [30, 44]. Talking about potential core zones of a digital permeation, systematic networking and structural change, the focus quickly shifts to the segment of logistics: here a wide range of applications for digital technologies is seen, "*which allow organizations to increase efficiency*

H. Lager (✉) · A. Virgillito · T.-P. Buchberger
TU Dortmund University, Sozialforschungsstelle, Dortmund, Germany
e-mail: hendrik.lager@tu-dortmund.de

A. Virgillito
e-mail: alfredo.virgillito@tu-dortmund.de

T.-P. Buchberger
e-mail: tom-philipp.buchberger@tu-dortmund.de

M. Klumpp and C. Ruiner (eds.), *Digital Supply Chains and the Human Factor*, Lecture Notes in Logistics,
https://doi.org/10.1007/978-3-030-58430-6_3

and productivity, whilst also providing greater transparency and accuracy in the movement of goods" [32, p. 176]. According to literature, intralogistics with its diverse fields of activities like warehousing, order picking, packaging, etc. is actually facing a major change in existing logistic structures and processes due to the digital transformation [45, 51].

Digital technologies interfere with the existing socio-technical system of companies and change the conditions of work [20]. Hence, they open up new and diverse possibilities to shape Human-Machine-Interaction. In the debate about digitalization and Industry 4.0 the implementation of digital technologies often refers either to potentials for better working conditions or the loss of jobs through automation [7, 11, 13, 43, 49]. While scholars today are more and more critical regarding the disruptive job loss hypothesis [5, 6, 31, 50], the results regarding better work conditions vary [24].

Better working conditions can arise in two different areas: The first area is linked to a higher degree of self-organization and autonomy and refers to a more fulfilling work experience. The second area are ergonomic improvements for employees by substituting repetitive and/or physically hard work in logistics through hands-free order picking, vital tracking, lifting aids, exoskeletons or reducing hand movements by the use of smart glasses. Against this background, through the substitution of ergonomically disadvantageous tasks we often see in consequence a rise of "easier" tasks. Instead of doing one straining work task, now three easier tasks have to be done. In this context, we analyze how the new digital technologies and ergonomic improvements and an intensification of tasks that follows relate to each other: Are opportunities of the substitution of ergonomically disadvantageous tasks used to integrate more challenging and attractive work or do they give rise to easier tasks? Does the work intensification eliminate ergonomic improvements? Here, we do not only relate the discussion on ergonomics and the discussion on work intensification, but also aim to enrich both debates, which still lack valid empirical results in the context of logistics. Focusing on order picking, which often is characterized by standardization, pre-structuring and physical stress [18], this chapter is structured as follows: In Sect. 2, we provide a short overview over different technologies and specific disorders and illnesses that occur in order picking. Based on that, we present the debates on ergonomics and work intensification in the context of digitalization. In Sect. 3, we present four empirical cases of the BMBF project "Innolab – Hybrid Services in Logistics" to analyze the relation between digital technologies, ergonomic improvements and work intensification. Finally, in Sect. 4 we sum up and discuss our results and draw conclusions.

2 Digital Technologies, Ergonomics, and Work Intensification

2.1 Digital Technologies in Logistics

A strong permeation and diffusion of digitalization on all levels is predicted in logistics [42, 51]. Although digital technologies like Cyber-Physical-Systems (CPS) are associated with increasing automation and efficiency, especially in logistics some obstacles prevent a complete automation of processes. This can be seen, for example, in the still very high proportion of manual activities, such as order picking. Here, the requirements to cope with dynamic and flexibility are very strong and employees play an important role in successfully handling those [18].

Therefore, digitalization projects often focus on "finding" the items rather than automating the picking process itself. Pick-by-Vision systems use for example devices like data glasses or tablets that are connected to the warehouse system via Wi-Fi, from where they receive relevant information for the order picking process. The device provides this information to the worker carrying out order picking tasks. Pick-by-Vision devices display information like contract orders, sequences, number of pieces etc. and guide the employee. Some models have a built-in scanner (e.g. camera) and eliminate the necessity to use an additional handheld-device.

A second technology variant are Pick-by-Voice systems. Instead of a handheld scanner and packing lists, the picker only needs a headset and a terminal when working with Pick-by-Voice. The terminal is connected to the warehouse management system and in most cases attached to the belt of the picker. The warehouse management system communicates the respective orders to him or her by radio. The system first calls a shelf. Once the employee has arrived at the shelf, the article and the quantity thereof are voiced. Once the warehouse worker has picked and loaded these, she or he confirms the process verbally. Then the next order is announced.

Thirdly, monorail systems, where the items are delivered to the picker, eliminate travel times of the worker and are more efficient in terms of picks per hour than the Picker-to-Part systems. Part-to-Picker systems, i.e. monorail systems, lack in flexibility regarding warehouse planning.

Eventually the working steps are simplified by the digital technologies and hence logistics is characterized by a large percentage of temporary work and a high level of low skilled work [3, 23]. In this chapter we focus on order picking. Such logistical activities often show a high standardization and pre-structuring and are often physically stressing (e.g. lifting up and loading heavy items, moving long distances) and require disadvantageous body postures [18, 48]. It is not surprising that pickers are often incapacitated by musculoskeletal disorders. In response, more and more companies develop ergonomic tools for order picking, as musculoskeletal diseases account for over 52% of all work-related illnesses in the European Union [17, p. 696]. Hands-free order picking and less travel time are often seen as ergonomically advantageous and are used to sell new technologies.

In general, according to data from Germany's largest health insurance company (AOK), the proportion of mental illness in all cases of incapacity to work has risen sharply [36, p. 447ff.]: It increased by 40% since 2008 and is responsible for about one in ten incapacities to work in 2018. However, most cases of incapacity to work in Germany are still due to respiratory diseases (just under a quarter), although these only account for a good 14% of all days of incapacity to work. Musculoskeletal diseases, on the other hand, are in first place in terms of sick days, which also account for the highest proportion of incapacity to work cases and days in the warehouse management sector. Within this sector, a quarter of the missing days and about 20% of the cases are due to musculoskeletal disorders [37]. This puts the warehouse management sector among the top ten occupational groups with the highest number of absences due to musculoskeletal diseases [36].

A second important factor is stress. Recent studies [2] report numbers for Germany as high as 80% of surveyed works council members indicate rising work intensity. Sick days due to mental disorders have more than doubled since 1994 from 1314 days to 2766 days per 1000 workers by 2012 [9]. While they are still a relatively small factor in comparison to musculoskeletal diseases, mental disorders are on the rise and with an increase in work intensity they are prone to become much more relevant.

Against this background, the implementation of digital technologies influences work in order picking systems. Here, we focus on ergonomic effects on the one side and potential work intensification on the other side. As the base for our further argumentation and our empirical findings, we show the corresponding state of research of the two lines of research which have not been connected to each other very often (see [18].

2.2 *Digital Technologies and Ergonomics*

Closely linked to the debate about digital technologies are changes in the Human-Machine Interaction and their consequences for work systems [20]. Here, the ergonomics of digital technologies and the shaping of work against this background is an important issue. Ergonomics as part of work research is a

> scientific discipline concerned with the understanding of interactions among humans and other elements of a system, and the profession that applies theory, principles, data, and methods to design in order to optimize human wellbeing and overall system performance. [21]

The aim of an ergonomic design of work systems is to optimize the workload, to avoid negative effects and instead to foster facilitating effects [1, 47]. The underlying idea is, that human performance, if it is not impaired, improves the effectiveness and efficiency of the system and thus might contribute to realize the goal of an ergonomic design of work systems. Hence, the human being should be regarded as the main factor and as an integral part of the system, its workflow and work environment. The field of ergonomics is divided into product and process ergonomics with different subareas like physical ergonomics, cognitive ergonomics and organizational ergonomics [1, 39, 42].

- *Physical ergonomics* deals with aspects like anatomy, anthropometry, physiology and biomechanics and the shaping of workplaces in order to avoid impairments of posture and movement.
- *Cognitive ergonomics* deals with human performance and human processing of information at the human-machine interface (e.g. perception, attention, memory, decision, motoric actions). It aims to shape and optimize the cooperation and interaction between human and technical system to enhance the efficiency of the whole human-machine system as well as the satisfaction and well-being of users.
- *Organizational ergonomics* focuses especially on work process related organizational aspects like division of labor, working times, teamwork etc.

Besides or across these three different fields of ergonomics mentioned above, there is also software-ergonomics. It deals with the adaptation of a software to the (mental and physical) characteristics, knowledge, behavior and habits of a user to ensure its usability. Ergonomics-related aspects in the context of technology and digitalization are also usability and user experience. *Usability* can be understood as the *"extent to which a system, product or service can be used by specified users to achieve specified goals with effectiveness, efficiency and satisfaction in a specified context of use"* [22]. In contrast to this, *user-experience* is defined as the *"users´ perceptions and responses that result from the use and/or anticipated use of a system, product or service"* (cf. ibid).

Many studies dealing with ergonomics in the context of Industry 4.0 and digitalization address software ergonomics with usability and user experience of digital technologies and interfaces [4] or ergonomic requirements in the context of collaborative robots [34, 41]. Surprisingly, the amount of laboratory studies or experiments seems quite high, so many studies do not (or only to a limited extent) take place in or consider the reality of companies with their complex conditions and challenges. The debate of ergonomics and digital technologies lacks long-term studies in the operational experience of companies. Grosse et al. [17, 18] criticize that, although order picking has a particularly high amount of human work and a high proportion of manual processes, a human-oriented and ergonomic design of work has

> often been ignored in management-oriented research (i.e. the production, operations and logistics management literature). Instead, researchers have focused on developing decision support models that help to achieve economic performance goals such as low cost or low order throughput times in the past [...], without paying attention to either the impact of system design on the human-in-the-system or the human influences on system performance. [18, p. 1261]

With regard to the implementation of digital technologies improvements in ergonomics and potentials to support employees, to reduce physical and/or psychological strain and to foster cognitive and senso-motoric capabilities are highlighted [14, 42]. In logistics, several stationary and mobile or wearable digital assistance systems already exist and are established in logistic companies.

Pick-by-Vision systems have recently become more sophisticated and comfortable [33]. Data glasses for example display relevant information dynamically and directly in the employees' field of vision.

Compared to paper- or tablet-based processes AR-technologies can support palletizing tasks and reduce palletizing errors [29]. Such logistic digital assistance systems providing packing information and instructions can reduce the risk of overloading (e.g. caused by damaging or forgetting articles), lower mental strain and frustration, and might have a positive impact on the motivation, satisfaction and performance of employees as well as on sickness rates [33]. Exoskeletons can support strenuous physical movements (e.g. lifting and picking heavy things) and reduce physical strain. In this context, Kinne et al. [26] show ergonomic potentials of using passive exoskeletons in palletizing activities where the palletizing task itself, the workload and task difficulty was rated lower by the test persons. Thus, the ergonomics of digital assistance systems highly depend on their specific usability and user experience, which influence the employees' rating of ergonomics and technology acceptance.

Despite these mostly positive reports and findings concerning the relationship of digital technologies and ergonomics in logistic work systems and especially in order picking, the following aspect should not be forgotten: usually, the main objectives of those digital assistance systems are, first of all achieving maximum picking performance, minimizing the error-ratio (e.g., wrong quantity or article type) and boosting the learning curve [33]. Hence, ergonomic improvements seem to be of a lower importance. Digital technologies can substitute physical demanding activities in logistics, have the potential to improve ergonomics and make work more interesting and varying. Nevertheless, some empirical findings show that in the course of implementing digital technologies an intensification of work often takes place.

This gap is also reflected in the current debate on the digitalization of work. Here, different scenarios of industrial and logistic work are discussed [19]. A positive scenario predicts rising employment effects, high-order tasks and qualifications, extended autonomy and all in all "*better jobs – jobs that at every level would be enriched by an information technology*" [52, p. 159]. A contrary, rather dystopian scenario assumes the emergence of negative effects that substitute and devalue industrial and logistic work especially on the low-skilled-level with a high degree of routine and restrict the autonomy of employees [13]. Work systems have a high degree of division of work and digital technologies like Cyber-Physical Systems (CPS) substitute human decision-making and crucial control functions of processes and work. This might also be called "*digital taylorism*" ([10, p. 15 ff., 46]). A third scenario group combines positive and negative effects and argues that a polarization of work occurs. Partial substitution processes and an erosion of work on the middle-level of qualifications take place and a shift to more complex tasks with high-order qualifications on the one side and a shift to easier tasks with low-order qualification on the other side might happen [15]. Hence, according to Goos and Manning [15], a differentiation of work into "*lousy and lovely jobs*" emerges. The concrete realization of this ideal-typical scenario depends on many aspects like the specific intention of the use of technology, market and customer requirements, wage levels or path-dependencies.

2.3 *Digital Technologies and Work Intensification*

Work intensity and intensification is not a new topic. Numerous studies have already shed light on it (for an overview see e.g. [40]). High work intensity is defined as the unbalanced ratio of the work result to be achieved (quantity/requirements) to the effort (time/resources). According to Ahlers [2] work intensification is the subjective perception of increasing work demands. Work intensification does not only appear in the form of increased workloads but can also be encountered as more complex performance requirements or more multi-tasking.

Reasons for work intensification can be manifold. Changing working conditions such as higher customer and market orientation, project work, mobile work or constant availability are all associated with higher work intensity [28]. Furthermore, staff shortages, technological innovations and changes in work organization, e.g. lean management, results-oriented work or target agreements [16] can cause work intensification.

High work intensity is a widespread phenomenon: More than 50% of the employees state that they work under strong deadline and performance pressure [8, 12]. Ahlers [2] conducted multiple analyses with the WSI Works Council Survey of 2018, a representative survey of ca. 2300 works council members in the private sector. 80% of the works council members report an increase in work intensity over the last two years in their respective companies. This varies with the specific employee group: 90% of skilled workers are affected, but only 44% of unskilled workers. 77% of the works council members think that work intensification has led to an increase in health problems, 68% think that the working atmosphere has deteriorated as a result. They see the main reason in thin staffing, which is not necessarily a consequence of illness or high order levels (which are also contributing factors) but is often described as normal and a rather conscious management decision. In addition, it is reported that vacancies are very difficult to fill. "Bad" management is often cited as a reason for unfavorable working conditions, such as permanent staff shortages, which in turn cause absences due to illness and result in declining work satisfaction.

Meyer et al. [38] conclude in their study that the introduction of new technologies is associated with an increase in work intensity but is also associated with opportunities and risks for both the work situation and the well-being of employees. Ittermann et al. [25, p. 160] show, on the basis of representative survey data, that in logistics there can be – and in comparison with other industries also above average – positive trends in the sense of workload reduction through digitalization (see Fig. 1). In their overview "The Intensification of Work" Paškvan and Kubicek conclude that

> [i]n many organizations, downsizing has been seen as a quick move to reduce costs, but the associated improvements are probably only temporary. In the long run, employees have more work to do, causing work intensification [...], stress-related illness and occupational accidents [...]. [40, p. 38]

Relating the results concerning the debate on digital technologies and ergonomics with the debate on digital technologies and work intensification, the following contradiction appears: On the one side, digital technologies have a high potential to

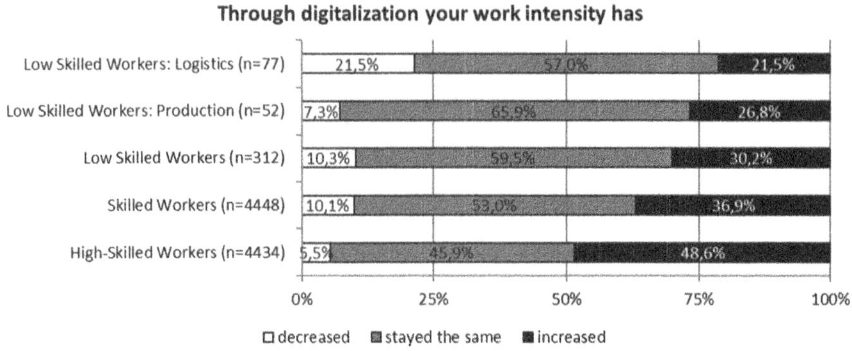

Fig. 1 Digitalization and work intensity. Source Ittermann et al. [25]

support employees, reduce physical and/or psychological strain and improve work ergonomics. On the other side, due to progressing digitalization, the work intensity has increased. This leads to the assumption that the intensification of work eliminates potential improvements in work ergonomics. Furthermore, emerging opportunities for the integration of more challenging and attractive tasks, due to the substitution of ergonomically disadvantageous tasks, might not be realized in companies. Against this background, we present our empirical findings in the following section to examine how digital technologies, ergonomic improvements and work intensification relate to each other in order picking. We also investigate whether opportunities of the substitution of ergonomically disadvantageous tasks are used to integrate more challenging and attractive work, whether they give rise to easier tasks and whether the work intensification eliminates ergonomic improvements.

3 Empirical Cases

The empirical cases our argumentation is based on stem from a research project called "Innolab – Hybrid Services in Logistics"[1] at the TU Dortmund and the Fraunhofer Institute for Material Flow and Logistics (IML) Dortmund. We conducted case studies including qualitative interviews, observations/warehouse tours, and the analysis of documents and websites in different companies, especially in order picking. For the evaluation of the interviews we followed a qualitative content analysis according to Mayring [35]. For this article, we selected four case studies where we analyzed 20 qualitative interviews with managers, works council members and employees. The interviews with the managers and works council members lasted from 90 to 120 min. Topics were consequences of the digitalization, ergonomics, and implementation of Industry 4.0-solutions (i.e. assistance systems). Interviews with employees were conducted during warehouse tours and lasted only a few minutes.

[1] https://www.innovationlab-logistics.com/

Table 1 Overview of empirical cases and interviewees

	Industry	Size	Interview Partners			Digitalization Solution		
			Management	Works Council	Employees	Pick-by-Vision	Pick-by-Voice	Monorail
1	Tools	600	1	2	1	X	–	–
2	Automotive	1000	3	2	1	X	X	–
3	Furniture	2200	2	2	2	X	-	X
4	Food	700	–	2	2	–	X	–

They consisted mainly of pointed questions regarding their own work experience and their use of the assistance systems. We chose the companies to present an overview of the different circumstances under which order picking is performed. From a tool manufacturing company in the steel sector with a small warehouse, to a local retail distribution center, an international distribution center, and an automotive supply warehouse, typical cases for order picking with varying plant sizes were chosen (see Table 1).

Case 1: Tool Manufacturer Warehouse
The company produces different high-quality tools (e.g. spanners, screwdrivers, torque tools and tester) with more than 600 employees. In the order picking area of the company work 10 pickers per shift. A Pick-by-Vision solution was implemented. Here, smart glasses are used in order picking processes. This case not only shows ergonomic effects and potentials of work intensification associated with a digital assistance system, but it also illustrates problems why the technology is not used by the employees in their daily work.

Case 2: Automotive Supplier Warehouse
The company has more than 1,000 employees and is working in contract logistics offering full-service-logistics and the material supply for assembly. As a service partner of the automotive industry almost all of its services (e.g. material supply of the vehicle final assembly, assembly of chassis modules, production of tank modules) focus on a single costumer. Order picking of products for production and assembling according to just-in-sequence and executing outsourced assembling processes for its customer are the core competences of the company. In different buildings different technologies are used in order picking to minimize error-rates: Pick-by-Vision and Pick-by-Voice .

Case 3: Furniture Retail International Distribution Center
Another case is a distribution center for an international furniture retail company. In the distribution center there are 2,200 workers employed (1,600 directly at the company and 600 via service partners). Work in the warehouses is done with forklifts or pallet trucks. Additionally, there is a monorail system where parts are delivered to the picker.

Case 4: Food Retail Local Distribution Centre
The company is active in food retailing. In the visited warehouse, articles from the dry and frozen product ranges are picked according to the principle "person to goods".

The employees in the warehouses work with a Pick-by-Voice system, which has been in use for about five years at the time of the survey, and use forklifts. A total of 700 order pickers are employed at the location, 80% of them are unskilled or lateral entrants from other professions.

In the next subsections we highlight how each digital solution is implemented in the respective firms and what consequences for ergonomics and work intensification arise.

3.1 Pick-by-Vision

Case 1: Tool Manufacturer Warehouses

The observed tool manufacturer uses a smart glasses Pick-by-Vision solution. The main purpose of implementing this Pick-by-Vision system was to enhance the efficiency in order picking, generate time savings in scan processes and increase the picking rate. This was planned because of the spatial restrictions of the warehouse: The order volume made an additional picker necessary, but the warehouse was too small to accommodate another picker with a palette truck.

Ergonomic improvements have not been addressed and have not been a crucial reason for implementing this technology. Even though the works council members mention some potential positive effects like reducing premature wear of employees by hands-free order picking, less physical bending, and fewer movements to reach back and overhead. Furthermore, an employee reports that the glasses improve the presentation of information.

> Well, I like the glasses. I prefer the glasses. Because it's just more comfortable on the eye. We've just got some parts which you have to search. You simply have the advantage that you don't have to go back to the terminal to look again: What was that again? Or something like that. I think that's quite pleasant, to see it straight out of the eye. (Employee order picking)

Overall, however, the company and most of the employees are very disappointed with the smart glasses. Before implementing the Pick-by-Vision system many employees doubted potential positive ergonomic effects. The smart glasses cause headaches and earaches due to the weight of the glasses. Due to this, many employees do not work with the glasses but with the old order picking technologies (tablets and handheld scanners). According to a works council member, against the background of potential physical strain (e.g. headaches, earaches, eye problems and fatigue), rest and break times for the users are important to prevent rising error rates in order picking that might occur in consequence of wearing the smart glasses over a long time and cause fatigue and psychological strain.

Additional problems arose with the use of the integrated camera for scanning labels and barcodes. The Pick-by-Vision solution had problems with old labels at storage locations that were difficult to scan. Hence, a time consuming and costly replacement process of a large number of labels was necessary to increase legibility of barcodes. The warehouse is characterized by narrow alleys causing jam problems for the order picking vehicles. Little space between the racks impedes and prevents the use of large labels and barcodes, sometimes there is also only a small distance between the individual barcodes themselves making scan processes for the employees with the glasses difficult and uncomfortable. Furthermore, the camera of the glasses is rigidly installed and does not correspond to the actual viewpoint of the employees. They have to perform specific head movements and postures to aim directly and precisely onto the barcode to be able to successfully scan it. An employee says, if he wants to scan a barcode, he has to force himself sometimes to *"look past on the left so I can hit the right"* (Employee warehouse). This affects ergonomics negatively because employees need to perform unnatural and unusual movements and physical bending. The result is a high frequency of errors because wrong labels of storage locations might be scanned. This can cause or strengthen time pressure and physical stress. Because of this, employees usually do not use the Pick-by-Vision glasses or only one function, namely the presentation of picking information. They rather continue using the old handheld scanning devices because they are easier to use and more comfortable.

The ERP-System provides the employees with orders, their priorities and their sequence of processing. "*It is also pre-set. The employee has no influence on what he gets. It's set up in advance. [...] And then he's guided through the warehouse*" (Employee warehouse). On the condition that the Pick-by-Vision system would work, it could facilitate training processes for new and unskilled employees: "*Because the work steps are so beautifully broken down means that only what is necessary, is displayed*" (Head of logistics). All in all, the scope of action and decision remains relatively small or "*is equally zero*" (Head of logistics) and did not improve due to the implementation of the Pick-by-Vision system—even if it would work. The main reason for the smart glasses was work intensification and ergonomic benefits came not into fruition. In the end, the glasses were rejected because of their ergonomic disadvantages.

Case 2: Automotive Supplier Warehouse

In order picking the automotive supplier uses a Pick-by-Vision solution via a tablet. Usually the order picking process with the tablet is organized as follows: The employee drives with an order picking vehicle through the alleys of the warehouse and the installed tablet on the vehicle displays the picking order consisting of a male or female first name and a number of pieces. The names correspond to specific storage compartments and replace the complex and error-prone handling of different numbers at the compartments. The storage compartments are ordered alphabetically per male and female names. The tablet visualizes the right storage compartment in form of the relevant name and also shows the relevant number of pieces to be picked for the sequence. After scanning the barcode of the compartment with a handheld scanner, the connector system on the vehicle shows an exact position and orientation for the pieces in the production sequence. A new scan shows, if the order picking has been carried out correctly by the employee, who then confirms this on the tablet. The order of the sequence is structured and automated by the transport control system and the Pick-by-Vision system. The usability of the tablets is very good and does not require specific technological knowledge, a "digital affinity" is sufficient. "*That's not a menu you have there. Instead you really see where you have to go and a confirmation. And then you scan something. And the scanning confirms it*" (Works council deputy chairman).

In the course of the implementation of the tablet, the error rate in order picking significantly decreased, but the throughput did not increase significantly. While the order picking work was already tayloristic and structured before, due to the Pick-by-Vision system via tablet, work in order picking became more tayloristic, i.e. much easier, more structured and time-bound. Due to this extended standardization of the work process, the scope of action and decision for the employees became highly restricted. Against this background, a rotation of work tasks is practiced to enhance the variety for employees in sequence order picking with the Pick-by-Vision system at least on a low level.

> We just want people to do a kind of group work. [...] And we simply assume that it doesn't happen, that the stronger person takes the best jobs, but that they organize themselves in their group and then practically get the orders. Everyone still does everything, provided that he or she is physically able to do it. (Head of business unit logistic)

Especially the works council refers to a huge control potential of processes and work in order picking, enabled and enhanced by the Pick-by-Vision system and the transport control system: "*We are transparent. We are completely transparent for [...] [the OEM]*" (Works council deputy chairman). Due to the combination of transport control system and Pick-by-Vision (as well as Pick-by-Voice) sequence errors "*can be traced back to the tip of the nose*" (Works council chairman). Although the interview partners did not mention mental stress, the potential of an increased traceability and transparency of individual performance and errors seems to us to be omnipresent and increases time and performance pressure for the employees and the pressure to comply with the defined error rates. Furthermore, optimization processes were conducted to optimize routes and travel times of the employees in order picking. "*Also for ergonomic reasons the employee should move, but also not walk too much during the day*" (Head of business unit logistic). However, the employees are now order picking more in the course of this optimization, which might overcompensate positive ergonomic effects.

Case 3: Furniture Retail International Distribution Center

In the case of the international distribution center the basic picking routine via tablet is the same as in the case before. The tablets and the accompanying Pick-by-Vision system have long been introduced and widely accepted. Although the tablet increases work intensity, it is under the supervision and control of the works council. They assure that the data generated by the tablets is not used to measure individual worker performances. A newer addition to tablet guided picking is a scan glove which makes the additional work step of confirmation of the picking via handheld scanner obsolete. Here the acceptance is quite good because the freedom of how to use the scan glove is high: The pickers can either use the glove, use a cuff if they do not want to wear a glove, or not use the new scanning tool at all. In the end, most workers are using the glove, some use the cuff and only very few stick with the old handheld scanner. The hands-free picking process is widely perceived as more comfortable. The aim of the scan glove was to allow for a higher number of pickings per shift. Ergonomic considerations were not taken into account and, as in the previous case, it is highly debatable if the ergonomic advantages are not eaten by work intensification in the long run.

To sum up, and in agreement with Grosse et al. [18], our Pick-by-Vision cases highlight that ergonomic reasons are either not considered at all. They seem to serve merely as a pretended reason for the legitimation to implement digital technologies, which aim to enhance productivity, reduce error-rates and gain more control and transparency of order picking processes. Rather, the empirical findings presented above show that more productivity goes hand in hand with higher work intensification. Ergonomic improvements that emerged where mostly not intended and are probably not enough to offset higher work intensity.

3.2 Pick-by-Voice

Case 2: Automotive Supplier Warehouse

The automotive supplier also uses a Pick-by-Voice system. The order picking process is structured via Just-in-Time and Just-in-Sequence. The system consists of a headset, a microphone, a headphone and a small PC. Due to the implementation of Pick-by-Voice technology, work in order picking became more tayloristic, i.e. simpler, more structured, monotonous, repetitive, and time-bound. Accordingly, the scope of action and decision for the employees have been reduced by using the headsets because the employee is very strictly guided by voice. The training time of the Pick-by-Voice headset for new or unskilled employees is very low. Processes of de-qualification are observable: now, employees require even less training to fulfill order-picking tasks.

Due to the implementation of the Pick-by-Voice system the error picking rate considerably decreased.

> Well, you have to say that this Pick-by-Voice does not necessarily make the employee faster in his work, but he is [...] just as fast. [...] But we've had an enormously high level of quality with it. [...] So for every million picks, we make about three mistakes. (Head of business unit logistic)

But in the course of analyzing and optimizing the routes and travel times of the employees, new potentials for rationalization emerged that have been used to intensify order picking. "*It's increasing. The more travel times are eliminated, the more picking rises*" (Works council). Here, such optimizations are possible because the Pick-by-Voice system allows the recording of moving profiles: "*They do not only see the route they are walking, they also see the time*" (Works council).

Due to analyzing and optimizing routes and travel times, optimizing sequence order picking and improving error rates by the Pick-by-Voice system, a part of the staff was reduced. "*We no longer need so many quality assurance [QA] personnel.*

Well, we've really had to give up two positions in every shift at QA, I think. There are six jobs broken off" (Works council). The face-to-face-interaction between QA-employee and order picking employee in this case ceased and was substituted by a standardization and technological control mechanisms reducing error rates in order picking.

From an ergonomic point of view, the Pick-by-Voice solution enables a hands-free order picking and hence facilitates order picking work by providing only relevant information that are required to fulfil the sequence. The employees can adjust the speed and acoustic signal or voice of the Pick-by-Voice headset. Due to the reduction of information and complexity as well as to the significantly decreased risk, that employees make mistakes in order picking potential overstraining in some cases could successfully be prevented.

Nevertheless, many employees have to cope with increasing psychological strain, despite the reduction of complexity, technological support and instructions of the Pick-by-Voice system. Some employees report a higher strain caused by the necessity to pay attention at any time wearing the Pick-by-Voice headset and listen to its instructions which might be perceived by the employees as a permanent acoustic irritation. In doing so, the employees need to keep an overview of their working surroundings to prevent accidents, for example with vehicles driving around. Although the headsets only have a speaker on one side, so that the order pickers are not completely shielded or isolated from their environment and its events, they need to stay alert all the time. This is aggravated by a huge degree of structuring, monotony and repeatability of order picking with the Pick-by-Voice headset. "*If you hear seven and a half hours: 'Go to point X. Extract [...] Please confirm.' If you hear that 200 times a day, you'll snap. Right?*" (Works council chairman). In case of malfunctions of the Pick-by-Voice system mental/cognitive stress might occur: "*And then malfunctions, right? Because the Pick-by-Voice doesn't run smoothly. Then you have to repeat it a thousand times. You're going crazy*" (Works council deputy chairman).

Another cause of increasing psychological stress is that every pick in the sequence order picking can be assigned to the corresponding logistical employee via his personnel number. According to a works council member, order picking in sequencing in combination with Pick-by-Voice allows for a huge control potential of work. Indeed, according to the interview partners an evaluation for performance and behavior control is possible, but they point out that it is not conducted in the company. The works council is implementing a works agreement that contains guidelines and information on how to deal with the introduction of new technological systems and data privacy requirements. Despite these negative effects according to a works council member, the acceptance of the employees of the Pick-by-Voice solution is quite high.

Case 4: Food Retail Local Distribution Center

In the case of the local distribution center, the Pick-by-Voice solution has a substantial effect on the physical and psychological well-being of the employees. On the one hand, the physical work increases. In this case, the picking volume has increased by 50% (in comparison to picking with paper lists). The ergonomic improvements resulting from hands-free order picking are thus countered by an increased picking volume and, as a result, a higher total lifting weight per shift. The reduction of "idle times" (like walking to the office to get the next picking list) is an additional factor. Although the work council attempted to counteract the work intensification by increasing the recovery time from five to ten percent, employees repeatedly pointed out the lack of breaks. This is summarized by a works council member as follows:

> The work itself, especially in the picking area, has been reduced to the most stressful activities. Because all the things you used to have, such as: I'm going to the office, what should I do next are no longer necessary. Because I have everything right here. And the proportion of this so-called picking, that is, this grasping and lifting, is of course considerably increased. That's physical, yes, as I said, the work has become harder, not easier. (Works council member)

The wearing comfort of the Pick-by-Voice solution also affects physical well-being. The headsets are perceived as uncomfortable by some of the staff and have led to headaches. However, it has improved since the system was introduced because the pickers have been equipped with the latest generation of headsets.

According to the employees, the psychological stress has also increased considerably. Many people find processing the announcements of the Pick-by-Voice system cognitively challenging.

"So, the biggest problem we have is the psychological pressure. Like I said, we get the headset on the head and they bombard us with numbers." (Works council member).

Another stress factor was, especially in the beginning, the disturbance of the system. These included Wi-Fi failures, unclear announcements, and not understood confirmations of the picker by the system. However, due to improvements by the employer, these disturbances have been reduced to a minimum.

The issue that was most often raised, apart from the aspect of increasing work, was "communication". On the one side, workers complain about psychological problems resulting from the fact that they have to communicate almost exclusively with a "monotonous computer voice" during their entire working hours.

> Because just this stupid wearing [of the headset] and this getting used to it, to hear a voice. Some people say 'Hey, I repeat that at night, right?'. Because all they're saying is a number, a check digit. She says, 'Take three, take three'. You say: 'Okay, okay'. Or: 'confirm' or whatever. And some even repeat it at home. (Works council member)

On the other side, as already mentioned in the previous section, the possibility of compensation is limited, as communication with colleagues is much less frequently due to the Pick-by-Voice system.

> [The communication is] decreasing. There are still some who can articulate, but that is not much anymore. They used to communicate more with each other. That's over. (Works council member)

By contrast, the limited possibilities for action are judged much more ambivalently. Thus, most of the employees have no problems with the "externally controlled navigation" through the warehouse. Some are even happy about the decreasing responsibility. Only a few workers prefer to pick on their own.

A last aspect concerns the transparency of work performance, which is also indirectly related to the Pick-by-Voice system. Due to the coupling of this form of paperless picking with the warehouse management system, the work of the employee can be tracked much better, because the picker must confirm each step of the picking process (naming the check digit on the respective shelf, confirming each individual picking process). This creates a feeling of permanent supervision among the employees and additionally reduces the willingness to communicate with colleagues or otherwise take a short break. It is not even decisive whether their data is used by the employer. According to the works council, this company only checks certain data if a picker's performance is noticeably poor over several days and weeks, which is related to the strong limitations by the "General Data Protection Regulation" (GDPR) in Germany.

To sum it up: Pick-by-Voice increases the physical and psychological stress of the employees. On the one side, there is an increased picking volume and a higher total weight per shift, an increase in cognitive effort ("bombardment with numbers"), in the isolation of the individual picker and in the transparency of their work performance. On the other side, there is a lack of compensation possibilities. The work of the employee is reduced to the most stressful activities ("pure picking"), because "more restful activities" are no longer necessary. Furthermore, communication with other employees is only possible to a limited extent. In addition, changes that (also) benefit the pickers (e.g. increase in recovery time, more reliable Wi-Fi network, and new generation of headsets) do not seem to have improved the physical and psychological well-being much.

3.3 *Electric Monorail*

Case 3: Furniture Retail International Distribution Center
The furniture warehouse uses two distinct systems in their warehouse. One is a traditional Picker-to-Part system which works with pallet trucks and forklifts. The other is comprised of an electric monorail which is directly connected to a fully automated small parts warehouse. The processing of outgoing and incoming goods is operated by a different worker. Of special interest in that warehouse are the electric monorail system as the most advanced one:

> Because EMS is the thing where most [...] is done technologically. They also have their own KPI dashboard, where they can see exactly how it works? [...]. So there is actually the greatest software support. (Deputy warehouse manager)

The technologically most advanced station proved to be the most ergonomically disadvantageous. The electric monorail system is connected to an automated small parts warehouse and transports goods to the picker where he or she commissions into a station. Workers need to bend down very low when putting the parts into the station which is even aggravated as the first parts to put there are normally the heaviest ones for which they have to bend down the most.

> And the employee stands all the time and then just has to pack up the goods that come by [...]. And there we have, and I want to say that, we definitely save order pickers. (Works council chairman)

While these rail systems are responsible for a rise in productivity, workers feel less in control of their work and through the work at the station they feel isolated (to a degree) from their co-workers as they work alone in these stations and any co-worker is at least ten meters away. Also, the work intensity is very high because there is always the next order coming. The workers found several ergonomic improvements themselves and had to insist on their implementation because of a skeptical management. One was to equip the station of the electric monorail with rubber mats. This made standing and kneeling on a step in front of the pallet in the station more comfortable.

4 Conclusion and Outlook

All in all, the presented cases seem to suggest, that all the talk about ergonomic approaches seems to be more of a fig leaf than a wide-spread strategy to solve one of most pressing problems in logistics in Germany today: labor shortage.

Most solutions presented here aim to raise the number of picks per shift, lower error-rates, diminish training time, and eliminate idle times. The foremost solution to this are various assistance systems and advanced software systems in the background. These solutions typically cut away tasks like getting a new order list to pick and planning a route through the warehouse. These tasks get replaced with more picks. The consequences are a higher work intensity and progressing taylorization of tasks, leading to the desired results: More picks per shift without the need to increase the number of employees and a further simplification of work tasks, thereby reducing the needed training time. Hence, we see a mostly digital taylorism scenario in order picking.

While in the short run this strategy seems to alleviate the labor shortage in the logistics sector by increasing the workload, first problems are already occurring. Because it is ignoring long-term consequences, this approach needs a worker surplus on the labor market. The rising work intensification should result in higher absenteeism due to illness and accidents in the long run. Which in turn leads to a higher turnover in the workforce which is counteracted by the shorter training times due to the increased tayloristic organization of work. For this to function, the labor market needs to be filled with job seekers. Otherwise the workers in that segment will seek out more attractive work, because the short-term solution makes this kind of work highly unattractive: It is boring, dull, and not well paid. Technology acceptance will be low due to the constrained scope of action and decision and the resulting low self-efficacy [27]. Since the system is built in its core on the exchangeability of the employees, the employees will not feel appreciated very much. Corporate loyalty will be low and for very minor increases in pay they will change employers. In short, exchangeability of workers adds bargaining power to the party which has the rarer commodity: Workforce or workplace.

Adding the aging of the German society into account, the labor shortage problem will increase twofold: Firstly, there are less new workers entering the labor market and secondly, the older workforce will have even more problems with the increasing workload. Since the demographic change is a widely known phenomenon by now, it is somewhat surprising that long-term strategies which incorporate ergonomic viewpoints are still relatively scarce and health prevention, corporate loyalty and corporate culture are still overlooked. Often mitigating suggestions (e.g. job rotation, increase in rest periods) come from works councils and have to be incorporated against employers' resistance.

In the light of the Corona pandemic of early 2020 two more contradictory factors have emerged. On the one hand, the increased demand for home delivery of goods instead of buying in a shop creates even more job opportunities in order picking and, as such, should elevate workers bargaining power. On the other hand, an economic recession seems to be nearly certain. Since order picking jobs are mostly low skilled, workers from other industries may seek jobs in order picking, thus changing the power balance yet again.

Logistics and ergonomics research is in dire need of more long-time studies under real working conditions. Even the few survey results which are available for logistics work (see Fig. 1) are based on relatively low numbers of cases.

References

1. Adler M, Hermann H J, Koldehoff M, Meuser V, Scheuer S, Müller-Arnecke H, Windel A, Bleyer T (2010) Ergonomiekompendium. Anwendung ergonomischer Regeln und Prüfung der Gebrauchstauglichkeit von Produkten. Online: BAuA. https://www.baua.de/de/Publikationen/Fachbeitraege/F2116-2.pdf?__blob=publicationFile&v=4. 13 April 2020
2. Ahlers E (2020) Arbeitsintensivierung in den Betrieben. Problemdeutungen und Handlungsfelder von etriebsräten. WSI Mitteilungen 73(1):29–37
3. Ahlstrand R, Been W, Boethius S, Corchado S, Fernández N, Gautié J, Green A, Iléssy M, Keune M, Koene B, Latniak E, Makó C, Martín F, Marthieu C, Payon N, Perez C, Postels D, Rehnströn F, Wright S, Warhust C (2018) Virtuous circles between innovations, job quality and employment in Europe? Case study evidence from the manufacturing sector, private and public service sector. Quality of jobs and Innovation generated Employment outcomes (QuInnE). Work package 6: Qualitative Analysis
4. Ahram T, Falcão C (2017) Advances in usability and user experience. In: Proceedings of the AHFE 2017 International conference on usability and user experience. Springer, Switzerland
5. Arntz M, Terry G, Zierahn U (2016) The risk of automation for jobs in OECD countries: a comparative analysis. OECD Social, Employment and Migration Working Papers No. 189
6. Arntz M, Terry G, Zierahn U (2019) Digitalization and the future of work: macroeconomic consequences. ZEW Discussion Paper No. 19-024
7. Autor DH (2015) Why are there still so many jobs? The history and future of workplace automation. J Econ Perspect 29(3):3–30
8. BAuA (2019) Arbeitswelt im Wandel. Zahlen – Daten – Fakten. Online: https://www.baua.de/DE/Angebote/Publikationen/Praxis/A100.pdf?__blob=publicationFile&v=6. 14 April 2020
9. BKK Dachverband (2013) BKK Gesundheitsreport 2013. Berlin

10. Brown P, Ashton D, Lauder H (2010) Skills are not enough: the globalisation of knowledge and the future UK economy. UK Commission for Employment and Skills (UKCES) Praxis 4
11. Brynjolfsson E, McAfee A (2014) The second machine age. Work, progress, and prosperity in a time of brilliant technologies. Norton & Company, New York
12. DGB-Index Gute Arbeit (2015) Der Report 2015. Wie die Beschäftigten die Arbeitsbedingungen in Deutschland beurteilen. Online: https://index-gute-arbeit.dgb.de/++co++bf0decf0-942a-11e5-82bf-52540023ef1a. 14 April 2020
13. Frey C B, Osborne M A (2013) The future of employment: how susceptible are jobs to computerisation? Oxford Martin School Working Paper 18
14. Georg A, Guhlemann K (2020) Arbeitsschutz und individuelle Gesundheitskompetenz. Perspektiven der Prävention von Arbeitsintensivierung in der „Arbeit 4.0". WSI Mitteilungen 73(1):63–70
15. Goos M, Manning A (2017) Lousy and lovely jobs: the rising polarization of work in Britain. Rev Econ Statistics 89(1):118–133
16. Green F (2001) It's been a hard day's night: the concentration and intensification of work in late twentieth-century Britain. Br J Ind Relat 39(1):53–80
17. Grosse EH, Glock CH, Jaber MY, Neumann WP (2015) Incorporating human factors in order picking planning models: framework and research opportunities. Int J Prod Res 53(3):695–717
18. Grosse EH, Glock CH, Neumann WP (2016) Human factors in order picking: a content analysis of the literature. Int J Prod Res 55(5):1260–1276
19. Hirsch-Kreinsen H (2016) Digitization of industrial work: development paths and prospects. J Labour Market Res 49:1–14
20. Hirsch-Kreinsen H, Ittermann P, Niehaus J (eds) (2018) Digitalisierung industrieller Arbeit. Die Vision Industrie 4.0 und ihre sozialen Herausforderungen, 2nd edn. Nomos/edition sigma, Baden-Baden
21. IEA (International Ergonomics Association) (2020) What is ergonomics? Online: https://iea.cc/what-is-ergonomics/. 14 April 2020
22. ISO (International Organization for Standardization) (2018) ISO 9241–11 Ergonomics of human-system interaction. Usability: definitions and concepts. Online: https://www.iso.org/obp/ui/#iso:std:iso:9241:-11:ed-2:v1:en. 15 April 2020
23. Ittermann P, Eisenmann M (2018) Digitalisierung von Einfacharbeit in Produktion und Logistik. In: Hirsch-Kreinsen H, Karacic A (eds) Logistikarbeit in der digitalen Wertschöpfung. Perspektiven und Herausforderungen für Arbeit durch technologische Entwicklungen. FGW. Düsseldorf, pp 57–76
24. Ittermann P, Niehaus J (2018) Industrie 4.0 und Wandel von Industriearbeit - revisted. Forschungsstand und Trendbestimmungen. In: Hirsch-Kreinsen H, Ittermann P, Niehaus J (eds) Digitalisierung industrieller Arbeit. Die Vision Industrie 4.0 und ihre sozialen Herausforderungen, 2nd ed. Nomos/edition sigma, Baden-Baden, pp 33–60
25. Ittermann P, Ortmann U, Virgillito A, Walker EM (2019) Hat die Digitalisierung disruptive Folgen für Einfacharbeit? Kritische Reflexion und empirische Befunde aus Produktion und Logistik. Industrielle Beziehungen 26(2):150–168
26. Kinne S, Kretschmer V, Bednorz N (2020) Palletising support in intralogistics: the effect of a passive exoskeleton on workload and task difficulty considering handling and comfort. In: Proceedings of the 2nd international conference on human systems engineering and design (IHSED2019), pp 273–279
27. Klumpp M, Hesenius M, Meyer O, Ruiner C, Gruhn V (2019) Production logistics and human-computer interaction. State-of-the-art, challenges, requirements for the future. Int J Adv Manuf Technol 105:3691–3709
28. Kratzer N (2020) Arbeitsintensität und Arbeitsintensivierung. WSI Mitteilungen 73(1):3–10
29. Kretschmer V, Plewan T, Rinkenauer G, Maettig B (2018) smart palletisation: cognitive ergonomics in augmented reality based palletising. In: Proceedings of the 1st international conference on intelligent human system integration (IHSI 2018), pp 355–360
30. Lager H (2020) Anpassungsfähigkeit in Zeiten der Digitalisierung. Springer VS Verlag für Sozialwissenschaften, Wiesbaden

31. Lehmann E, Wilhelm D (2018) Digitalisierung, Disruption und Corporate Entrepreneurship. In: Keuper F, Schomann M, Sikora LI, Wassef R (eds) Disruption und Transformation Management. Springer Gabler, Wiesbaden, pp 239–266
32. Mahroof K (2018) A human-centric perspective exploring the readiness towards smart warehousing. Int J Inf Manage 45:176–190
33. Mättig B, Kretschmer V (2019) Einsatz digitaler Assistenzsysteme in der Logistik 4.0. In: Vogel-Heuser B, Bauernhansl T, ten Hompel M (eds) Handbuch Industrie 4.0 Bd.3. Springer, Berlin, pp 5–35
34. Maurice P, Padois V, Measson Y, Bidaud P (2017) Human-oriented design of collaborative robots. Int J Ind Ergon 57:88–102
35. Mayring P (2000) Qualitative content analysis. Forum Qual Social Res 1(2). Online: https://www.qualitative-research.net/index.php/fqs/article/view/1089/2385#gcit. 14 April 2020
36. Meyer M, Maisuradze M, Schenkel A (2019) Krankheitsbedingte Fehlzeiten in der deutschen Wirtschaft im Jahr 2018 – Überblick. In: Badura B, Ducki A, Schröder H, Klose J, Meyer M (eds) Fehlzeiten-Report 2019. Digitalisierung - gesundes Arbeiten ermöglichen. Springer, Berlin, pp 413–477
37. Meyer M, Maisuradze M, Schenkel A (2019) Krankheitsbedingte Fehlzeiten nach Branchen im Jahr 2018. In: Badura B, Ducki A, Schröder H, Klose J, Meyer M (eds) Fehlzeiten-Report 2019. Digitalisierung - gesundes Arbeiten ermöglichen. Springer, Berlin, pp 479–717
38. Meyer S-C, Tisch A, Hünefeld L (2019) Arbeitsintensivierung und Handlungsspielraum in digitalisierten Arbeitswelten – Herausforderung für das Wohlbefinden von Beschäftigten? Industrielle Beziehungen 26(2):207–231
39. Mühlstedt J (2016) Grundlagen virtueller Ergonomie. In: Bullinger-Hoffmann A C, Mühlstedt J (eds) Homo sapiens digitalis - Virtuelle Ergonomie und digitale Menschmodelle. Springer, Berlin, pp 7–39
40. Paškvan M, Kubicek B (2017) The intensification of work. In: Korunka C, Kubicek B (eds) Job demands in a changing world of work. Impact on workers' health and performance and implications for research and practice. Springer International Publishing, Basel, pp 25–43
41. Pearce M, Mutlu B, Shah J, Radwin R (2018) Optimizing makespan and ergonomics in integrating collaborative robots into manufacturing processes. IEE Trans Autom Sci Eng 15(4):1772–1784
42. Rinkenauer G, Kreutzfeld M, Renker J, Plewan T (2018) Kognitive Ergonomie in der Logistik. In: Kopp R, Ittermann P (eds) Konzeptionelle Perspektiven von Arbeit in der digitalisierten Logistik (Soziologisches Arbeitspapier 55), pp 35–55
43. Rüßmann M, Lorenz M, Gerbert P, Waldner M, Justus J, Engel P, Harnisch M (2015) Industry 4.0. The future of productivity and growth in manufacturing industries. The Boston Consulting Group. Online: https://www.zvw.de/media.media.72e472fb-1698-4a15-8858-344351c8902f.original.pdf. 14 April 2020
44. Spath D, Ganschar O, Gerlach S, Hämmerle M, Krause T, Schlund S (2013) Produktionsarbeit der Zukunft - Industrie 4.0. Stuttgart
45. ten Hompel M, Henke M (2017) Logistik 4.0 – Ein Ausblick auf die Planung und das Management der zukünftigen Logistik vor dem Hintergrund der vierten industriellen Revolution. In: Vogel-Heuser B, Bauernhansl T, ten Hompel M (eds) Handbuch Industrie 4.0. Band 4. Springer, Berlin, pp 249–259
46. The Economist (2015) Digital Taylorism. 12. September 2015, p 63
47. Tosi F (2020) Design for ergonomics. Springer series in design and innovation, vol 2. Springer Nature, Switzerland
48. Walch MD (2011) Belastungsermittlung in der Kommissionierung vor dem Hintergrund einer alternsgerechten Arbeitsgestaltung der Intralogistik. Lehrstuhl für Fördertechnik Materialfluss Logistik der Technischen Universität München, München. Online: https://mediatum.ub.tum.de/doc/1187877/fml_20131230_70_export.pdf. 10 Mar 2020
49. Walwei U (2016) Digitalization and structural labour market problems: the case of Germany. ILO Research Paper No. 17. Online: https://www.ilo.org/wcmsp5/groups/public/---dgreports/---inst/documents/publication/wcms_522355.pdf. 10 April 2020

50. Wienzek T, Virgillito A (2018) Leise Innovation, nicht radikaler Umbruch. Die Einführung einer Industrie-4.0-Lösung bei einem Möbelhersteller – ein Fallbeispiel. WSI Mitteilungen 71(3):204–210
51. Winkler H, Zinsmeister L (2019) Trends in digitalization of intralogistics and the critical success factors of its implementation. Br J Oper Prod Manage 16(3):537–549
52. Zuboff S (1988) In the age of the smart machine: the future of work and power. Basic Books, New York

Hendrik Lager is researcher at the Social Research Centre at the TU Dortmund University, Faculty of Social Sciences. He studied sociology at the University of Göttingen and received his Ph.D. from the Faculty of Business and Economics at the TU Dortmund University. His research focuses on digitalization of work, leadership in the digital transformation and the adaptability of companies.

Alfredo Virgillito is researcher at the Social Research Centre at the TU Dortmund University, Faculty of Social Sciences. He studied social sciences at the University of Duisburg-Essen. His research focuses on digitalization of work, industrial relations and participation.

Tom-Philipp Buchberger works at the Social Research Centre at the TU Dortmund University and is currently studying sociology at the University of Duisburg-Essen for his master's degree.

Implementation

Technology Bundles in Maintenance: Ambivalences Between Virtuality and Assistance

Martin Eisenmann and Nick Große

Abstract The provision of physical resources, their interconnection by the internet and the creation of data as well as its exchange and transferring via corresponding virtual communication, increases complexity. The increase of complexity based on the growing data structure, various functionalities of digital technologies and diversity of the range of work designate the challenges in industrial maintenance. Based on a complex data structures and its virtual use—or otherwise: *because of* the complex data situation—there are optimal conditions given for the application of innovative technologies within industrial companies. In this context, digital technologies provide one or several functionalities (e.g. "virtuality" or "assistance") and for those functionalities certain technologies are typical (e.g. Digital Twins or assistance systems, such as tablets or data glasses). At the same time, the implementation of these technologies includes consequences for human labor (e.g. in industrial maintenance). For this reason, the two modes of functionalities mentioned are discussed, using the example of the so-called "Digital Twin" and its combination with assistance systems in industrial maintenance.

1 Introduction: Two Functionalities of Technologies in Smart Maintenance

A central aspect of the vision Industry 4.0 [1] is the holistic interconnection of human workers, machines, objects and their corresponding information and communication technology to a Cyber-Physical System (CPS) [21, 33]. Those systems consist of sensors and actuators, which are interconnected via internet, and are able to communicate with each other [19, 22]. Due to the increasing use of sensors and systems for data acquisition as well as the systematic use of large data sets available [38], there are new opportunities but also challenges for industrial applications, which contribute to extensive changes from a socio-technical point of view (see Sect. 3).

M. Eisenmann (✉) · N. Große
TU Dortmund University, Dortmund, Germany
e-mail: martin.eisenmann@tu-dortmund.de

M. Klumpp and C. Ruiner (eds.), *Digital Supply Chains and the Human Factor*, Lecture Notes in Logistics,
https://doi.org/10.1007/978-3-030-58430-6_4

An example for a technological innovation is the versatile use of Digital Twins (DT) in production or industrial services and maintenance. The use of a DT has various use cases and benefits: First, for simulating processes in factories due to its ability to simulate realistic processes (e.g. services of maintenance) along the entire value chain (function of virtuality). Reasons for the use of a DT and its virtuality functions can be seen in the resulting transparency [1], whereas from a critical point of view, the possibility of increasing distances or gaps between reality and virtuality are present [1]. Distance refers to the often imprecise virtual representation of production and process realities. Second, DTs are used for Condition Monitoring (CM) [21], which describes the continuous and real-time condition tracking of technical assets [1]. The combination of CM data and digital assistance systems [21, 33] allows the operationalization of a DT for operational activities. In terms of applicability in maintenance and production scenarios, this could be achieved via prepared context-sensitive information that is displayed for example via tablets, (digital) checklists, or Augmented Reality applications (function of assistance). On the one hand, referring to the assistance function, human knowledge and participation remain relevant to compete with the increasing complexity. On the other hand, due to the assistance systems, new opportunities for complexity reduction, control potentials and adjusted task assignments for operational activities become apparent. Both, the assistance and the virtuality function are within the scope in this chapter. The functionalities, which have different effects or consequences for human work and knowledge structures, can occur either separately or simultaneously. In the case of the DT, different consequences can be observed, so that they appear inconsistent, which this chapter describes as ambivalence. Based on the below sketched assumptions (Sect. 2) regarding the virtuality and assistance function, this article answers the following research question:

What are the consequences for labor regarding the functionalities of Digital Twins and assistance systems in industrial maintenance?

This contribution begins by analyzing the function of digital technologies (see Sect. 3). In addition, the DT, which combines both of the focused functions (assistance and virtuality), is defined. Afterwards, potentials and fields of application (use cases) for DTs are listed in the context of industrial maintenance and how these are integrated into operational processes (Sect. 3.1). Furthermore, important aspects of both functionalities are listed in contrast to each other with regard to industrial maintenance. In terms of the consequences for human labor associated with the multifunctional use of technology bundles in maintenance processes, a classification into conflictual and beneficial consequences is made (Sect. 3.2). Based on this classification, the ambivalences[1] of the consequences are extracted (Sect. 3.3).

[1]In the context of this article, ambivalence of consequences means that some consequences are not always clearly conflictual or beneficial and identifiable. They are often contradictory and occur in some maintenance cases simultaneously and in other cases not at the same time (see Sect. 3.3).

2 Underlying Assumptions and Aim of the Chapter

A DT is the virtual representation of assets, products, machines or intelligent service processes based on a connected data structure [1], which often appears in a combination with assistance systems. In this article, such a combination is called *technology bundle*. The mentioned database is fundamental for the *virtuality* of the DT, which manifests itself in forms of simulations or models of production and maintenance processes. But, the growing database generates an increasing complexity [33], which requires the mastery of data about service or work processes and thus *assistance*. Concerning DTs, it seems there is a non-ambiguous or an ambivalent relation between both functions (assistance and/or virtuality, see Sect. 3) and they can occur either separately or simultaneously.

Based on the mentioned problem, this chapter is based on two assumptions: First, the growing complexity [1, 33], which is based on the increased amount of data and use of new technologies, lead to new opportunities in terms of virtuality and assistance for industrial applications (for example by supporting operational activities or via an improved process simulation). Second, we assume that different combinations of digital technologies [21] have different effects on human labor, which appear in an ambivalent way.

First, the aim of this article is to answer the above mentioned research question. Furthermore, it gives a closer look to the potentials and challenges of technology bundles using the example of the DT. The DT is characterized and defined as well. Second, the contribution analyzes the interconnection of virtuality and assistance by listing up conflictual and beneficial consequences for human work and performance in the field of industrial maintenance. The analysis focuses on the interdependencies between the usage of technology and its human operator. Regarding those interdependencies between multifunctional usage of technology bundles and maintenance labor, this chapter works out the ambivalences of those consequences.

3 A Socio-technical Perspective on Smart Maintenance

This chapter follows a socio-technical point of view on changes and innovation. As an instrument of analysis [22], the socio-technical approach focuses not only on specific technical innovations but also on the whole system (e.g. a company), and therefore on the changes in the organizational and personal sub-systems and especially on their interdependencies [19]. Hirsch-Kreinsen [19] notes that due to the functionalities of the technical sub-system restrictions are set, but that interdependent properties affect the functionalities of the technological subsystem. Furthermore,, Ittermann et al. [22] points out, systems interact closely with their environments. In another study that is also sharing a socio-technical basic understanding in terms of the innovation, [21] state four functionalities of digital technologies (see Fig. 1). The authors assume that implementing “new technological systems do not go along with a consistent pattern

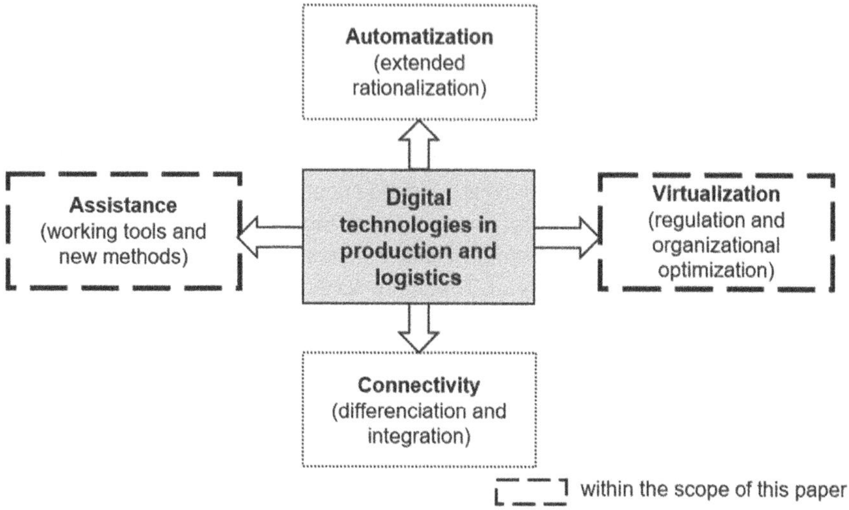

Fig. 1 Core functionalities of digital technologies [21]

of production and logistic work" [21]. Therefore, a "differentiation of functionalities of digital technologies" [21, p. 37] could be useful. This differentiation provides the basis for further analysis and implications in the design of digital work. The illustration below (see Fig. 1) demonstrates the previously mentioned four functionalities named Automation, Virtuality, Assistance and Connectivity. Especially the two functionalities Assistance and Virtuality play an important role in this chapter and will be considered using the example of DT and assistance systems in maintenance.

According to Ittermann and Falkenberg [21], automation technologies function as an adaptive media of the rationality, most importantly on the progressing substitution of operational and strategic work and processes (for example robotic systems or autonomous transport and storage units). The functionality of connectivity or rather the interconnection of machines, humans, products, tools and other technical elements is worth mentioning, which simultaneously allows a fast and partly cross-company communication. Examples for these functionalities in terms of digital technologies are cloud-services, external or inter-company platforms and the Internet of Things. This enables a differentiation of previous limits and working models (also for the maintenance) and leads to an "innovative integration of processes and procedures" [21p. 37ff]. The last two functionalities of digital technologies are within the scope of this paper and thus receive more attention:

On the one hand the virtuality function is justified by the connected "engineering—and planning system(s) and advanced process monitoring system(s), which are able to preprocess a large number of parameters" [21]. In accordance with the authors, these are also called organization technologies. Examples for such technologies are the DT itself but also Enterprise Resource Planning (ERP) systems, controlling or shop floor management systems. Due to the processing of plenty of production data and their conditions, digital technologies fulfill the function of virtualizing real processes for regulation and control, or they fulfill the simulation of production and its accompanying resources and also the workforces [21], which until now were realized through the organization and personal themselves. The virtuality function goes along with the standardization of (working) processes and their optimization. In conjunction with the virtuality function methods of (conventional) data acquisition, processing and evaluation like: knowledge discovery methods in terms of industrial databases, Data Mining [38], which are used for Machine Learning correlations and offering optimization potentials (for example for predictive maintenance scenarios) [21], are of particular importance in this context. The individual components are virtually connected via the use of mobile devices. At this point, a close conjunction with assistance systems, which is addressed by the second functionality of digital technologies focused in this contribution, is apparent. Hence, the term of technology bundles is used in this chapter.

In addition to that, according to Ittermann and Falkenberg [21], especially the above-mentioned mobile devices or assistance systems are functioning ergonomically supporting, guiding or assisting. These technologies change parts of their functions in response to human behavior, such as: physical characteristics, qualification, responsibility, powers of activity, or level of knowledge. Adaptive or digital assistance systems provide an "ergonomic centered adjustment on specific working conditions" [21, p. 43ff], which for example are "hands-free" activities, situation specific and context-sensitive information-presentation or the reduction of "physical, mental or social impairment" [21]. The design of the man-to-machine interaction and their tasks vary widely, because from a critical point of view, the use of digital assistance systems is associated with "technological control, lack of scope for action and also intensified performance" [21]. Forms of assistance systems are for example data glasses combined with a virtual or augmented reality application, tablets or wearables [12], which allows the redesign of working processes, for example via remote instructions in the field of maintenance. Both functionalities of digital technologies can be found in industrial maintenance and its service processes. These central categories will be examined more closely in this section. The explored scenario is determined by the applicability of DT and assistance systems within the above-mentioned economic segment.

3.1 Virtual Assistance in Smart Maintenance Using the Example of the DT

To give a common understanding of the DT from a socio-technical point of view, an overview regarding definitions is introduced. In addition, the functionalities of the DT are described and projected onto the fields of application within the context of maintenance. Here, appropriate use cases with already existing examples are shown. The content is derived from a review of relevant literature. In the following, the potentials are highlighted as well as the challenges of using the DT as a part of the technology bundles. This is done from the point of view of virtuality.

3.1.1 Basic Facts About the DT and the Concretization of Subject of Consideration

The frequently used term DT has plenty of definitions such as being a "digital representation of a real entity" [18]. The term was defined for the first time by [15], who describes it as a "digital equivalent of a product" whose adaption followed in the year 2010 by NASA[2] [15, 26, 44, 46] .

In this context, terms like digital model or digital shadow are mentioned or merged [26, 34]. To provide clarity, a differentiation of these terms will be described below. Depending on the interaction between the real component and its virtual version, a digital shadow is characterized by an unidirectional way of communication flow from the physical component to its virtual one, whereas the DT enriches the functionalities of a digital shadow due to its bidirectional communication flow from the virtual component to its real pendant [26]. Apart from the DT and digital shadow, a digital model has no data exchange with its real component [26]. Due to the numerous amounts of definitions, a suggestion for an holistic overview of definitions DT is made in [47].

In order to differentiate between a static three-dimensional model such as a prototype, the DT considers the behavior along the whole product lifecycle [47] through its consistent link to its real component, including the design, production and also service stage, whereas the scope of interpretation of a prototype is only limited by the design stage [47]. For both, the quality of the model is depending on the processed data [47]. For visualizing the cause-effect relationships, suitable 3D-models, such as CAD-based replications, are essential for the DT [2].

A deeper look is set on the industrial maintenance, which interacts with the production in a coexistent way [34]. Both domains are interacting separately and aiming their own objectives without, which equals a silomentality [29]. Despite of the different views on DTs, which can be seen in the numerous amount of partially intersecting definitions, the consideration of the holistic product lifecycle makes the DT appropriate for interdisciplinary use such as in joint maintenance and production scenarios [34]. According to Tao et al. [47], there are design, production and service applications [47]. Keeping the purpose of this chapter in mind, which is the provision of services in the fields of operational maintenance scenarios, a deeper insight to the service application is given.

---

[2]According to NASA, "A Digital Twin is an integrated multi-physics, multi-scale, probabilistic simulation [...] that uses the best available physical models, sensor updates [...], to mirror the life of its flying twin.[...] In addition, [...] the digital twin integrates sensor data from the vehicle's [...] system." [46].

Functional Scope and Fields of Application of the DT

The previous section dealt with the different point of views in regard to the consideration of the DT. In a next step, the essential characteristics of the DT will be derived. Central characteristics within the DT are the illustration of a real-physical object and their conditions, which jointly enable the virtual representation of the systems' cause-effect relationships Anylogic [5].

According to their functional areas, the fields of applications can be summarized to [34, 37]:

- derivation of prescriptive maintenance actions,
- monitoring to reveal cause-effect relationships of influencing parameters and its appropriate visualization,
- decision support, increasing awareness in the sensitivity of the influence parameters,
- adjustment of the machine parameters for the increase of the overall equipment efficiency (OEE),
- synchronizing the maintenance and production planning,

The operational maintenance can benefit from the enrichment through dynamic virtual models due to its possibility to draw conclusion for prognosis and to identify indicators for the potential malfunctions of an asset [18]. The DT transforms maintenance from a once reactive to a now proactive discipline, for example through the analysis of the abrasion behavior or rather its remaining useful life [3]. Through the combination of condition monitoring and the functionalities of the DT, changes of conditions and its effects on the performance can be projected onto the whole system as well as onto individual components and thereupon be evaluated [24].

Challenges

An essential problem concerning the handling of maintenance relevant data for the data-driven maintenance is the existence of numerous data sources and its various forms of deposit (for example some part of knowledge is stored in the head of the employees) [34], which leads to a lack of retrievability (from external sources) and information transparency. Furthermore, a challenge can be seen, on the one hand, in the continuous illustration of the objects' behavior in the virtual environment, and on the other hand, the handling with lack of information [45]. For a transparent cause-effect loop, the information management has to ensure a consistent document basis without contradictions [45].

Data which are of high relevance and are used in a cross-domain context such as within the joint planning of production and maintenance [26] can be virtually displayed and, additionally, in case of changes in the production setting, returned to the real-physical component in forms of changes in the configuration of the system parameter. With the help of CM and DT the aim of a condition-based maintenance can be met to a certain extent [23]. The acquisition and the appropriate linkage of the data of the whole factory and its correct interpretation is a challenge [43].

Tao et al. summarize essential challenges related to the use of DT, which moreover must be met equally taking into consideration balanced cost-benefit ratios and comprise [47]:

- the lack of awareness in regard to its value contribution and its possible intervention in the employees' comfort zone,
- the lack in design of an appropriate and demand-oriented DT,
- the increase in numerous isolated solutions and its missing interfaces and the requirement, and
- the requirement concerning the robustness and safety of such systems.

All of these challenges and requirements must be met equally taking into consideration balanced cost-benefit ratios [26]. To get a preferably detailed representation of the situation in real-time, an appropriate development environment is required Anylogic [5]. Furthermore, an adequate information management is required to ensure consistency of the documents or the information contained therein [45].

Potentials

Due to the near real-time data processing and the interaction between the real component and its virtual equivalent, the DT is an extensive tool, which unfold its full potential through its functionality in the handling of big data [42]. Within the scenario of a CPS, the DT provides possible applications through its intermediary interacting function between the real and virtual layer [42]. Further detailed industrial fields of applications are mentioned in Tao et al. [47], which one can see that the aviation industry is a good pioneer in the applicability of the DT for maintenance purposes. Further highlighted application scenarios mentioned in Kritzinger [26] are seen in the automotive section in terms of calculating the potential failure rate and thus the appropriate time for initiating a maintenance task. The same use can be seen in the maritime scenario in terms of tracking the availability of a ship and deriving the appropriate maintenance intervals. An analogue use case mentioned in Kritzinger et al. [26] can be transferred to "Prognosis and Health Management (PHM)" on wind turbines [47].

A notable best practice of the DT for asset scenarios mentioned in Negri et al. [37] is a project called "i-Qui". The appropriate assignment of asset components within the whole system is the focus, which is demonstrated by type plates where for each of the plates an individual DT has been generated. Here, the DT aims at the 'digital itemization of all used asset components' [18].

In a mature development stage the DT may contribute to a risk reduction with a simultaneous increase in efficiency, quality, flexibility, and transparency [52]. The interplay between man and machine raises the question how both participants can learn from each other (mutual learning) [4, 34]. In the future, it might be interesting to decide whether and which planning decisions shall be transferred from the human to the machine and at what point does the human being still have the authority of decision-making [18]. Together with its behavior characteristics, the real physical system is displayed as well in the virtual system [45], so that—depending on the quality of the DT—causes of malfunctions can be derived, for example based on correlations and parameter settings.

Supported by the available data base of the DT, potential indicators for preventive actions and—depending on the completeness of the underlying information – more favorable work flow sequences can be derived [45]. According to the spectrum of implemented virtual elements and the design quality of its behavior, which can be improved by machine learning algorithms,[3] in intention to imitate the real behaviour of the system at best [47].

According to the quality or maturity of the DT, a potential use case for the operational maintenance is the assistance in learning, which contributes to reducing the effort for the tasks to be performed to a reasonable extent [18]. It allows to focus on alternatives and "What if?" scenarios, without testing these on the real component. Furthermore, it can contribute to an increase of awareness of cross-domain relationships, for example the above-mentioned joint planning tasks for production and maintenance [34]. In addition to that, the use of AR/VR devices and its corresponding immersive environment contributes in the exploration of scenarios, for example the effects caused by the change of parameters [47]. In combination with AR/VR-applications, the DT is considered as an enabler for remote maintenance scenarios [14].

3.1.2 Interim Conclusion: Suitability of DT for Virtual Assistance

Most of the use cases, almost independently from their application scenarios (maritime, aerospace, automotive or production), are subsumed under the fields of Prognosis and Health Management (PHM) purposefully to reveal potential cause-effect relationships between the parameter settings and the course of the system behavior. Referring to the virtuality function, the question of the amount and the level of detail of the projected information is essential, whereas the assistance function has to deal with the question to what extent the user has to be guided within the interaction of the DT.

Using the example of an AR scenario inside a factory where an employee is responsible for maintaining its corresponding assets, he or she will be able to get an insight into the total amount of data of the whole asset park, which leads to an information overflow and thus has to be reduced to the relevant quantity. Assistance could be achieved through a pre-processing done by machine learning algorithms with or without the perception of the user. Another important question is which functionality should be transparent and which should be executed automatically in the background (without the noticing of the employee) due to their importance or the potential vulnerability (sensitivity).

---

[3] For an explicit list of suitable algorithms, a reference is made to [47].

As the revised literature mentioned above confirms, there are plenty possibilities to support maintenance to reveal important information for the appropriate maintenance time. However, the sole implementation of artificial intelligence and technical devices does not allow transparency about the whole system and its corresponding processes. It is more likely a question about how and to what extent the initial purpose is served. A general statement about the ideal ratio of virtuality and assistance cannot be given for the whole spectrum of application scenarios consisting of an interplay of artificial and human intelligence.

The already mentioned and summarized potentials show that a high potential in the use of the DT can be seen in the derivation of cause-effect relationships in the operational maintenance [1, 25]. This leads to the question whether and how both above-mentioned functionalities can contribute to the prediction of appropriate maintenance intervals. Due to the applicability of VR and AR, the virtuality function may unfold its potential possibly. However, due to the reason that the use of the DT goes along with its underlying device, such as a technology bundle, the application potential can only be as good as the connected interface allows, which is why the above-mentioned statements should be treated with caution. Thus, the dependency on the user interfaces sets the limits of possibilities for the functionalities considered in this chapter.

3.2 *Consequences of Digitalized Maintenance: Between Assistance and Virtuality*

In this section, the functionalities of digital technologies, with a focus on maintenance and industrial services, will now be contrasted with regard to their interdependencies and ambivalent consequences for human work. In the context of this article, the term digital technology implies the combination of the digital assistance system and the DT as well as the associated use of data, which in our view can be described as a technology bundle. For this reason, both functions of this technology bundle will be analyzed with regard to their consequences for the work process in the economic sector of maintenance and industrial service. On the basis of those consequences, the conflictual as well as the beneficial aspects will be analyzed in order to show the ambivalences of the functions.

3.2.1 Assistance Function in Smart Maintenance: Consequences for Labor

There is no doubt that the implementation of digital technologies and their assisting function is associated with beneficial or advantageous effects for labor in industrial enterprises. Central elements of those effects for the economic sector of maintenance are listed below:

Support and complexity-reduction. The use of assistance systems as a "tool" [52] in an industrial contexts fulfils supporting and assisting function. Depending on the intended use of the technology, the authority over control, responsibility, and decision-making processes regarding the value chain and labor processes still lies with the industrial skilled worker, who can intervene in those processes independently and with guidance [22]. Niehaus [39] also sees assembly, logistics and maintenance as fields of application for assistance systems (e.g. smartphones, watches, data glasses, RFID bracelets, headsets, etc.). The purpose of the assistance systems are audiovisual preparation of work instructions, learning content or information as well as decision support. In addition to Ittermann and Falkenberg [21], who attribute a supporting, adaptive and guiding function to assistance systems for human activities, Kunath and Winkler [28] also state the complexity reducing component of assistance systems, which on the one hand has a supporting effect and on the other hand, in combination with the technology of the DT, constitutes the core of so-called adaptive assistance systems. This combination (in this article referred to as technology bundles) takes assistance systems to a new level, as they can learn from decisions made by professionals and pre-sort proposals for action for future decisions [28]. The pre-sort proposal of assistance systems are similar to modern recommender systems from e-commerce, which are now also used in industrial areas, e.g. in the form of data glasses [11].

Employment of basic qualified workers. The use of this employment group is another positive aspect to be mentioned in the scope of the assistance function regarding the high fluctuation of skilled workers in companies of the most industrial segments. As early as 2015, Acatech [1] points out that skilled workers with a basic qualification will receive work instructions by remote used assistance systems in the future. The study differentiates between so-called generalists, who are guided by a central maintenance authority, the specialists and data analysts, within the framework of defined work steps. This takes account of the lack of skilled workers, partly due to demographic factors in the industrial maintenance departments and in the industrial services sector. Maintenance and industrial services would thus become accessible to less qualified employment groups, which in turn would possibly lead to positive developments in the labor market. At the same time, there would be a shift in the focus of maintenance tasks towards the responsibility of information and communication technologies in combination with the mechanical components of industrial machines and plants [1].

Ergonomic improvement and flexibility. Ergonomic improvements are a big plus in terms of the assistance function of digital technologies. In this context, the socio-technical design criteria of digital work, such as adaptivity, complementarity and context sensitivity [22, p. 38], would result in the case of assistance systems as "aspects of an ergonomically oriented adaption of digital systems to specific working conditions and stresses, possibly systematic stress control or the automation of particularly stressful activities" [22]. With regard to the effects of rationalization, according to Ittermann and Falkenberg [21], higher efficiency and a reduced failure rate, but also an increase in decision-making ability and flexibility are particularly worthy of mention. The use of assistance systems goes along with a flexible work place and time and also with a spatial split [39]. This gives maintenance work more mobility and a higher degree of self-organization.

Accumulation of tacit knowledge. For the field of industrial skilled work, the above-mentioned increase in complexity (e.g. the rising data structures), according to Böhle [9, 10], also means that tacit knowledge is reproducing itself through the accumulation of practical experience, which goes far beyond routines and habits. It rather constitutes itself from the elements of activity such as feeling and develop a sense for technology and data, the presumption of a malfunction or the right idea for troubleshooting etc., and is therefore always embedded in practical action and a specific (working) context. The similar "[…] increase in the significance of tacit knowledge and experience-guided work with the rising degree of complexity and mediatization of technical systems" [41p. 347] is observable. It is confirmed that digital assistance systems have the quality of "expert or knowledge systems for maintenance activities" [21], which is also accompanied by the "enrichment and enhancement of work through on-the-job learnings and extended scope for action" [21].

With regard to the assistance function of digital technologies in maintenance and industrial services, however, in terms of their consequences for work, conflictual interdependencies are observable. Central aspects of these consequences will now be analyzed:

Increased control and lack of scope for action. "From a critical perspective, arguments of technological control, (and) lack of freedom of action" [21] can be observed. Shafto et al. [46] assumes in his study that the system's specifications on content and the underlying speed of work would, on the one hand, sanction processes of manual work at the operational level (e.g. through error messages). On the other hand, the implementation of assistance systems and the documentation of location, status or process data in industrial practice is associated with the simultaneous implementation of control and monitoring mechanisms. Those mechanisms observe or monitor performance, which consequently create enormous pressure on employees at the operational level [39].

For the economic sector of logistics, Niehaus [39] reasonably refers to the latent control potential that can manifest itself in tayloristic employment[4] structures and have a de-qualifying effect. De-qualification in this context means that parts of formerly holistic activities are broken out of task bundles. The remaining tasks stay more monotonous, isolated and fragmented so that they require less qualification than before [31].

Knowledge-and task-automation and increasing workload. Assistance systems automate partial activity elements [21]. Although there is no substitution of entire bundles of tasks, parts of maintenance activities are removed or automated from the activity spectrum, whose function can then be taken over by an assistance system, e.g.

[4]Lager [31] describes tayloristic employment as: "rigid, very finely structured division of work into many small, individual steps, social isolation of the employees." Furthermore, central elements of tayloristic structured organization are: standardization of operational labor, formalized and centralized as well as pronounced hierarchical structures.

decision-making or organizational tasks [28]. In particular, the removal of decision-making tasks, ordering activities, reduction of maintenance activities in the case of adapted maintenance intervals and automated warning functions that are communicated via assistance systems [13], are examples of the partial substitution of activities in maintenance [17]. Hirsch-Kreinsen [20] states that with increased automation and the intensive use of digital technologies (e.g. digital assistance systems) considering the increased complexity, failure potentials also increase. As a consequence, this means that no valid knowledge and experience about work and production processes could be built up. These failure potentials are found particularly in the difference between virtual and real processes (see Sect. 1), which means the loss of the ability to act, problem-solving and the accumulation of tacit knowledge in terms of industrial skilled work. With regard to assistance systems, this includes that on the basis of the system's prescriptions, competencies and the growth of tacit knowledge are prevented at the same time, since abstract and problem-solving oriented thinking, intuition, associative acting, etc. are shortened, because the paths to solutions are already predetermined. Ittermann and Falkenberg [21] also note an increased performance intensity associated with the use of assistance systems, and Niehaus [39] also confirms that assistance systems, in connection with the "threatening scenery" created by monitoring and control mechanisms, condense work performance at the operational level.

Abstraction of maintenance work and knowledge. Using the example of scheduling maintenance processes, Malsch [48] illustrated already a few decades ago that tacit knowledge in maintenance is in danger of being transformed or objectified by EDP systems into "planning knowledge", which the author also describes as a company-organized "process of data acquisition" [48]. This transformation process would therefore have two consequences: First, "objectification"—in other words an abstraction of knowledge structures and activities based on them; and second, the transformation increases the "opportunities for disposal and control" [48] for the management. It becomes clear that the described criticism of the technology has considerable parallels, both in the past and today, and must therefore be seen as a conflictual factor with regard to the assistance function of digital technologies. The "digitalization of knowledge into explicit information" [8] becomes also apparent in the context of DTs, which subsequently leads to a "formalization of knowledge" [8].

3.2.2 Virtualization Function in Smart Maintenance: Consequences for Labor

The second function of digital technologies focused here is the virtualization in the field of maintenance. In principle, technologies include a virtualization function providing "planning, control and monitoring procedures" [21] by using prepared data and information on production and service processes. This database is associated with technical, organizational and personnel elements and acts in part autonomously. The virtualization function also contains both beneficial and conflictual aspects with regard to its consequences for maintenance work and knowledge assets. Core elements of these consequences with a focus on the technology bundle (DT and assistance system) are:

Learning facilitation. Virtually presented content can be used to standardize and shorten learning phases and to train staff [30]. Especially for low-skilled and non-specialist personnel, virtual elements are used to prepare and deploy such employment groups for more demanding activities [39]. The opportunity of using 3D-based models of plants and machines, simulations are used for training and maintenance instructions, e.g. using VR (Virtual Reality) or AR (Augmented Reality) technologies. In particular AR is the "the needs-based supplementation of the field of perception with computer-supported content, (that) offers a promising solution" [36] and is found in the context of activity instructions. For example, data glasses, tablets or smartphones are also increasingly used in maintenance, since context-sensitive information is used via virtual channels to expand the perception of real processes. In contrast, VR technologies are more often found in the context of training scenarios [6]. With regard to the "provision of information from databases, knowledge management systems, manufacturing execution systems (MES) and enterprise resource planning systems (ERP)" [12], it should be mentioned that learning content, instructions (e.g. audio-visual and acoustic information) and commands are also transported via so-called wearables. These are mostly algorithmized information and data that form the basis of the DTs [16, 17].

Effects of rationalization: time and costs. Digital technologies that have a virtualization function, e.g. the DT, usually fulfill rationalization goals such as: synchronized division of labor, continuously integrated production processes as well as the reduction of unproductive working time and an increase in productivity through the acceleration of industrial processes [21]. Based on the results of the EU-funded project "IMPROVE",[5] in which an industrial alarm system was developed, rationalization effects go hand in hand with the implementation of the DT, which is integrated in the alarm system. Fullen and Pethig [13] attest DT in context of the project the increase of efficiency and the reduction of the failure rate as well as the reduction maintenance costs in the context of diagnosis and optimization of components on different plants and machines.

Illustration and transparency. The technology of the DT and its combination with assistance systems will be realized especially in industrial manufacturing and maintenance [16, 17]. A differentiation should be made between the representation of tangible objects, such as machines, plants, tools and products, and intangible objects, such as production or maintenance processes representing the status information of all objects via virtual or computer-supported models. As a result, not only the industrial assets, but also production and maintenance processes—and thus human work—can be illustrated and increase their visibility in the value creation process. The processes are becoming increasingly transparent. Condition monitoring, i.e. the measurement, recording and monitoring of operating parameters in real time, is a basic requirement for the virtual representation of plant fleets and maintenance processes. By processing and evaluating the data and parameters using stochastic algorithms, e.g.

[5]Link to the project IMPROVE: https://improve-vfof.eu/.

by the method of Data Mining [38], it is possible to predict the remaining service life of machines and their components as well as future-oriented failure and wear forecasts, which are implemented as part of software products in maintenance and service processes [27]. The combination and use of these technologies describes the strategy of predictive maintenance [49]. Predictive maintenance and in particular the recognition of breakdown patterns and corresponding forecasts enable, in addition to the knowledge gained from experience, a transparent decision-making basis for operational and strategic maintenance work.

The implementation of virtual technologies increasingly causes interactions in the operational structures, which can also lead to conflictual factors for the development of maintenance processes and labor. These central aspects are now being analyzed:

Difficult usage and build-up of tacit knowledge. In the course of the increase in complexity in production and production-related areas, such as maintenance, as well as the increased use of technology [1, 33], the base of data and information in maintenance, i.e. explicitly available knowledge, also increase [40]. This explicit knowledge appears in form of documentations, reports, drawings, operating and maintenance instructions, protocols, fault messages, etc. [50]. As outlined above, these explicit data sets will first be displayed digitally or virtually [16, 17]—this is what is meant by virtualization [21]. Furthermore, they are decentrally accessible for operational maintenance via the use of assistance systems. As described in Sect. 3.2.1, this leads to a possible deployment of basic-qualified skilled workers, in a critical perspective, however, this means that an excessive guidance and instruction of the staff by the data-systems would reduce the accumulation of implicit or tacit knowledge like: problem-solving competence, dealing with imponderables, the ability to think abstractly, associative thinking, creativity and other competences, that can be summarized under the label of subjective experience-guided abilities [10, 9, 41]. According to Malsch [48], the knowledge of such operational capabilities would be transfered into "planning knowledge" within the framework of cyclical phases. In such a perspective, the development of tacit knowledge and corresponding abilities to act would be considerably more difficult and consequently not used.

Simplifying and standardizing maintenance work. Ittermann and Falkenberg [21] note that with the virtualization function of digital technologies standardization of maintenance work appears. This could be seen as a result of strict process instructions and specified criteria. In the context of the virtual function of digital technologies, work is increasingly controlled and externally supervised. In the context of rational aims, the focus lies on optimization through the use of virtual elements, which at second glance are also related to the "tendency for standardization and simplification of processes" [21]. With regard to service technicians, [51] highlights the simplification of documentation and data management and the facilitation of service processes through the use of checklists. The degree to which this means a de-qualification of previously extensive and holistic maintenance and service activities remains to be examined and therefore offers scope for further research activities.

Gap between virtual and real process. Due to the increased level of virtuality, there is a risk of increasing functional and informational distance between the virtual and real processes (see Sect. 1). Hirsch-Kreinsen [19] refers in this context to monitoring activities in which the increased level of media or technology usage (e.g. through assistance systems) has led to a distance from the virtual and real production reality. The consequence would be that: "the informational feedback associated with vibrations, sounds, and smells, that many operators relied upon" [32] as well as the associated tacit knowledge and subjective-experiential skills and competencies [9, 10, 41] (e.g. associations, intuition, tact, dealing with unpredicted problems etc.) are reduced. The danger is conceivable, that due to the increased level of media, technology and data use, the virtual elements will distance (see Sect. 1) the knowledge and experience of the real process. In this concern the skills of service technicians require a fast and effective diagnostics ability of the failures and the competency of resolution, based on theoretical and technical expertise [41]. But those skills must be "linked to the knowledge based on experience and experience-led working skills" [41] and thus the knowledge is to a great extent based on experience and stays very important [41] for the everyday service work.

3.3 Findings

The analysis of the technology bundle and especially the analysis of the DT as well as the contrasted consequence for labor in the maintenance, based on the both functionalities (assistance and virtuality), are result of an extended literature review. On the one hand, this article provides different perspectives, definitions and limitations with regard to the DT and at the same time, it shows how the DT is related to mobile or adaptive assistance systems. Finally, following the exploration of various research sources, it appears that the combination of both technologies—of course in different configurations for each application—will be present in smart maintenance as a ready-to-use technology bundle. Wherever virtual content is transferred or visualized, the recipient often has a smart device or assistance system in use. At the same time, the contents on the assistance systems are digital and many of these contents have a virtual character. Subsequently, in Sect. 3.1 the conditions (e.g. real-time data and adequate information management as well as a suitable simulation environment) for a successful implementation of DTs were shown as well as the intended use. Those intentions mainly are: handling of large amounts of data, intermediate function, condition monitoring, displaying of cause-effect relationships as well as decision support in maintenance processes. In addition, this chapter offers a good insight into various application areas of the DT.

The two functionalities described above, i.e. assistance and virtualization, have been proven to be associated with work consequences in the industrial and maintenance sector. Some of the consequences can be evaluated positively with regard to the human-oriented criteria of labor and its design (adaptivity, complementarity, context sensitivity) [22], according to the Human-Machine Interaction:

- support and complexity reduction,
- employment of basic qualified workers,
- ergonomic improvement and flexibility,
- accumulation of tacit knowledge,
- learning facilitation,
- effects of rationalization: time and costs,
- illustration and transparency.

Nevertheless, conflictual consequences for maintenance labor can also be observed with an emphasis on the virtualization and assistance functions, which are rather disadvantageous to the human-oriented design criteria of labor:

- increased control and lack of scope for action,
- knowledge-and task-automation,
- increasing workload,
- abstraction of maintenance work and knowledge,
- simplifying and standardizing maintenance labor,
- difficult usage and build-up of tacit knowledge,
- gap between virtual and real process.

A closer look at the conflictual or beneficial consequences for labor in terms of the socio-technical design outlined above shows that some conflicts of goals become apparent. The functions of the technologies or technology bundles do not follow a linear process, as shown by the consequences for human labor, but demonstrates ambivalent development [20, 48]. Some of those ambivalences are listed up below:

(1) Concrete ambivalences can be seen, for example, in the highly divergent scope for action: On the one hand, assistance systems and the virtual content to be interpreted can be described as "expert or knowledge systems for maintenance activities" [21]. On the other hand, Pick-by-Voice systems for example, provide repetitive instructions to personnel in the logistics context. In the context of the assistance function, a differentiation is useful, which separates conflictual command systems and beneficial information systems [39].

(2) Ambivalence also can be observed in the handling of tacit knowledge or how it is differentiated depending on the deployment strategy and function of the technology bundles: Indeed, adaptive assistance systems and context-sensitive content have the effect of transporting knowledge and the independent availability of place, time and person. This in turn means that knowledge (e.g. problem-solving and search skills, abstraction skills, communicative skills) and the ability to act are reduced and devalued through these. However, the use of tablets, smartphones, laptops or software applications also integrates new skills into the "activity portfolio" of the maintenance staff, e.g. media skills, IT skills, data handling skills, etc., which in turn significantly enhances the range of action skills.

(3) A further contradiction in the work consequences lies between the increased transparency through simulations or models and the distance between the real process and virtual maintenance processes: The precise illustration of the DT

means that the scope of process changes can be better illustrated and surveyed and that interventions can even be made in the planning phase if necessary. Whereas, there is the danger of an increasing distance between real maintenance processes and the virtual images of these processes based on the higher level of mediatization in monitoring activities [19]. Although the industrial service or maintenance processes are thus mapped transparently and adequately, important information such as vibrations, odors, noises etc. could be misinterpreted and therefore lead to wrong decisions. It would also be more difficult to acquire tacit knowledge as a result of made experience. Nevertheless, Thomas [48] notes that "abstraction and intellectualization" of work seem compatible.

4 Conclusion: Complementary Relationship Between Human Work and Virtual Assistance

In conclusion, it can be stated, that because of the differentiated usability as well as the ambivalent consequences for work in the maintenance, a conflict of objectives between the two functions of digital technologies becomes apparent. To answer the above-mentioned research question, it is to be stated, that due to the different ways in which digital technologies function, the results of our literature review shows an ambivalent and contradictory configuration of the consequences for labor in the economic sector of industrial service and maintenance.

The outlined ambivalences allow the assumption that the assistance and virtualization function are closely related. Therefore, this article would also like to give the impetus for further development of the functionalities of digital technologies according to Ittermann and Falkenberg [21], for example in the form of a differentiation that could be called virtual assistance. The counterpart to this could be a functionality that is described as assisting virtuality, depending on which function dominates in practice and which degree of virtuality or assistance is appropriate.

Despite all virtuality and assistance as tools, which can also be attached with contradictions and will therefore hardly be linear in maintenance practice, the human worker remains the "performer of value creation" [7], who will increasingly depend on digital technologies, since they have a supporting, informing and complexity-reducing effect on maintenance processes in many cases. Following on from this, the importance of human work action in the context of establishing dynamic knowledge transfer in CPS is emphasized: "Although improvisation, intuition and creativity can be supported by artificial intelligence in case of unknown malfunctions or missing solution specifications, they can ultimately not be replaced and can only be achieved by humans" [35, p. 164]. This clearly shows the limits of virtualization and technical assistance and, accordingly, of the mediatization levels in production facilities and service processes. These limits must be kept in mind during the development and also during the implementation of new technologies as well as for the socio-technical design criteria, which must be taken into account when designing the

Human-Machine Interfaces: "adaptivity", "context sensitivity" and "complementarity" [22]. Considering the technological limits and the ambivalences described, it can be argued that the introduction of smart maintenance systems can only be successful if there is a balanced complementary relationship between the use of technology and data in combination with highly relevant experience and related knowledge as well as abilities of human workers.

Acknowledgements This contribution is part of the research activities of the Graduiertenkolleg (GRK 2193), funded by the German Research Foundation (DFG) (Project number: 276879186 GRK2193).

References

1. Acatech (2015) Smart Maintenance für Smart Factories: Mit intelligenter Instandhaltung die Industrie 4.0 vorantreiben. (acatech POSITION). Herbert Utz Verlag, München
2. Adler S, Masik S (2020) Der digitale Zwilling für virtuelle Fabrikplanung und-betrieb. In: Orsolits H, Lackner M (eds) Virtual reality und augmented reality in der digitalen produktion. Springer Fachmedien, Wiesbaden, pp 191–215
3. Aivaliotis P, Georgoulias K, Alexopoulos K (2019) Using digital twin for maintenance applications in manufacturing: state of the art and gap analysis. In: IEEE International conference on engineering, technology and innovation (ICE/OTMC), pp 1–5. https://doi.org/10.1109/ICE.2019.8792613
4. Ansari F, Hold P, Mayrofer W et al (2018) AUTODIDACT: introducing the concept of mutual lerning into a smart factory industry 4.0. In: Proceedings of 15th International conference on cognition and exploratory learning in digital AGE (CELDA 2018), pp 61–68
5. Anylogic (2020) An introdruction to digital twin development: whitepaper. https://www.anylogic.de/resources/white-papers/an-introduction-to-digital-twin-development/
6. Aromaa S, Väätänen A, Aaltonen I et al (2007) A model for gathering and sharing knowledge in maintenance work. In: Greef T de, Marasek K, Dittmar A et al (eds) Proceedings of the european conference on cognitive ergonomics 2015-ECCE '15. ACM Press, New York, USA, pp 1–8
7. Bauernhansl T (2014) Die Vierte Industrielle Revolution—Der Weg in ein wertschaffendes Produktionsparadigma. In: Bauernhansl T, ten Hompel M, Vogel-Heuser B (eds) Industrie 4.0 in Produktion, Automatisierung und Logistik, vol 46. Springer Fachmedien Wiesbaden, Wiesbaden, pp 5–35
8. Biedermann H (2016) Optimierung der Instandhaltungsstrategie durch datenanalytische Risikoklassifikation und Störungsprognose, pp 71–88
9. Böhle F (ed) (2017) Arbeit als Subjektivierendes Handeln: Handlungsfähigkeit bei Unwägbarkeiten und Ungewissheit. Springer VS, Wiesbaden
10. Böhle F (2017) Digitalisierung braucht Erfahrungswissen. https://denk-doch-mal.de/wp/fritz-boehle-digitalisierung-erfordert-erfahrungswissen/
11. Brandl P, Aschbacher H, Hösch S (2015) Mobiles Wissensmanagement in der Industrie 4.0. In: Weisbecker A, Burmester M, Schmidt A (eds) Mensch und Computer 2015. Workshopband. Wissenschaftsverlag, Stuttgart, pp 225–232
12. Evers M, Krzywdzinski M, Pfeiffer S (2019) Wearable computing im betrieb gestalten. Arbeit 28(1):3–27. https://doi.org/10.1515/arbeit-2019-0002

13. Fullen M, Pethig F (2014) Digital twins for industrial alarm management. In: visit–Industrial IoT—digital twin. Fraunhofer IOSB, pp 10–12
14. Geissbauer R, Schrauf S, Berttram P et al (2018) Digital factories 2020: shaping the future of manufacturing
15. Grieves M (2014) Digital twin: manufacturing excellence through virtual factory replication. Whitepaper 1:1–7
16. Güntner G (2018) Digital twins: Herausforderungen, Anwendungen und Nutzen von digitale Zwillingen in der Instandhaltung. In: Güntner G, Höller L (eds) Die digitale Transformation der Instandhaltung, pp 28–31
17. Güntner G (2018) Haben Ihre Anlagen schon eine IP-Adresse: Grundlagen und Anwendungsbeispiele für den Einsatz von IoT-Technologien in der Instandhaltugn. In: Güntner G, Höller L (eds) Die digitale Transformation der Instandhaltung, pp 32–35
18. Henke M, Heller T, Stich V (2019) Smart maintenance—Der Weg vom Status quo zur Zielvision (acatech STUDIE). (acatech STUDIE)
19. Hirsch-Kreinsen H (2014) Wandel von Produktionsarbeit—"Industrie 4.0". Soziologische Arbeitspapiere(38), pp 1–47
20. Hirsch-Kreinsen H (2015). In: Hirsch-Kreinsen H, Ittermann P, Niehaus J (eds) Digitalisierung industrieller Arbeit, 1st edn. Nomos Verlagsgesellschaft Baden-Baden, pp 10–31
21. Ittermann P, Falkenberg J (2019) Funktionsweisen digitaler Technologien und Szenarien digitalisierter Einfacharbeit. In: Hirsch-Kreinsen H, Ittermann P, Falkenberg (eds) Szenarien digitalisierter Einfacharbeit: Konzeptionelle Überlegungen und empirische Befunde aus Produktion und Logistik, 1st edn. Nomos Verlagsgesellschaft, Baden-Baden, pp 37–68
22. Ittermann P, Niehaus J, Hirsch-Kreinsen H, Dregger J, ten Hompel M (2016) Social Manufacturing and Logistics Gestaltung von Arbeit in der digitalen Produktion und Logistik. Arbeitspapier Nr. 47, Dortmund
23. Karbaum H (2017) Automatische Inspektion unbesichtigter Anlagen. In: Henke M (ed) Best practices for smart maintenance: instandhaltungsforum 2017. Tagungsband des 17. InstandhaltungsForums, pp 42–50
24. Kaur MJ, Mishra VP, Maheshwari P (2020) The convergence of digital twin, IoT, and machine learning: transforming data into action. In: Farsi M, Daneshkhah A, Hosseinian-Far A et al (eds) Digital, vol 55. Springer International Publishing, Cham, pp 3–17
25. Kleijn A (2018) Digitale Doppelgänger: Digitale Zwillinge machen den Lebenszyklus physikalischer Produkte transparent. Heise Magazine (online) iX 9/2018, pp 118
26. Kritzinger W, Karner M, Traar G et al (2018) Digital Twin in manufacturing: a categorical literature review and classification. IFAC-PapersOnLine 51(11):1016–1022. https://doi.org/10.1016/j.ifacol.2018.08.474
27. Kuhn T (2017) Digitaler Zwilling. Informatik Spektrum 40(5):440–444. https://doi.org/10.1007/s00287-017-1061-2
28. Kunath M, Winkler H (2019) Adaptive Assistenzsysteme zur Entscheidungsunterstützung für die dynamische Auftragsabwicklung: Konzeptionelle Überlegungen und Anwendungsszenarien unter Berücksichtigung des Digitalen Zwillings des Produktionssystems. In: Obermaier R (ed) Handbuch Industrie 4.0 und digitale Transformation, vol 23. Springer Fachmedien Wiesbaden, Wiesbaden, pp 269–294
29. Kurz M (2018) Koordination zwischen Instandhaltung und Produktion mittels Handelsmechanismus. Dissertation, RWTH Aachen; IIF—Institut für Industriekommunikation und Fachmedien GmbH
30. L. v. Schwerin M, Hofmann G (2020) Virtual reality in der Instandhaltung. https://www.instandhaltung.de/maintenance-in-der-industrie/virtual-reality-in-der-instandhaltung-306.html. Accessed 05 Feb 2020
31. Lager H (2020) Anpassungsfähigkeit in Zeiten der Digitalisierung. Springer Fachmedien Wiesbaden, Wiesbaden
32. Lee JD, Seppelt BD (2012) Human factors and ergonomics in automation design. In: Carayon P (ed) Handbook of human factors and ergonomics in health care and patient safety, 2nd edn. CRC Press, Boca Raton, FL, pp 1615–1642

33. Luft N (2013) Aufgabenbasierte Flexibilitätsbewertung von Produktionssystemen. Dissertation, TU Dortmund. Dortmund
34. Matyas K (2019) In: Instandhaltungslogistik: Qualität und Produktivität steigern, erweiterte. Praxisreihe Qualitätswissen. 7 edn Hanser, München
35. Meyer G, Knüppel K, Möhwald H, Nyhuis P (2014) Kompetenzorientiertes Störgrößenmanagement: Mitarbeiterkompetenzen sinnvoll in das Störgrößenmanagement integrieren. Prod Manage 19:15–18
36. Michalik A (2017) Augmented maintenance game. In: Henke M (ed) Best practices for smart maintenance: Instandhaltungsforum 2017. Tagungsband des 17. Instandhaltungs Forums, pp 81–89
37. Negri E, Fumagalli L, Macchi M (2017) A review of the roles of digital twin in cps-based production systems. Procedia Manuf 11:939–948. https://doi.org/10.1016/j.promfg.2017.07.198
38. Nemeth T, Bernerstätter R, Glawar R, Matyas K, Sihn W (2015) Instandhaltung 4.0: Sicherstellung von Produktqualität und Anlagenverfügbarkeit durch einen echtzeitbasierten Instandhaltungsleitstand. ZWF 110(9)
39. Niehaus J (2017) Mobile assistenzsysteme für industrie 4.0: Gestaltungsoptionen zwischen autonomie und kontrolle. FGW—Forschungsinstitut für gesellschaftlcihe Weiterentwicklung eV. 2017/09(04), pp 1–4
40. Nonaka I, Mader F, Takeuchi H (2012) Die organisation des Wissens: Wie japanische Unternehmen eine brachliegende Ressource nutzbar machen. Campus Verlag, Frankfurt/ New York
41. Pfeiffer S (2017) Subjektivierendes Arbeitshandeln im technischen Service. In: Böhle F (ed) Arbeit als Subjektivierendes Handeln: Handlungsfähigkeit bei Unwägbarkeiten und Ungewissheit. Springer VS, Wiesbaden, pp 347–360
42. Qi Q, Tao F (2018) Digital twin and big data towards smart manufacturing and industry 4.0: 360 degree comparison. IEEE Access 6:3585–3593. https://doi.org/10.1109/ACCESS.2018.2793265
43. Sauer O (2014) Über big data und die Optimierung von Anlagen mit Condition Monitoring und intelligenter Datenanalyse. https://www.iosb.fraunhofer.de/servlet/is/7942/Big%20Data%20Interview.pdf?command=downloadContent&filename=Big%20Data%20Interview.pdf
44. Schroeder GN, Steinmetz C, Pereira CE et al (2016) Digital TwinF data modeling with automationml and a communication methodology for data exchange. IFAC-PapersOnLine 49(30):12–17. https://doi.org/10.1016/j.ifacol.2016.11.115
45. Schüller A, Modersohn A, Kawohl M et al. (2019) Der Digitale Zwilling in der Prozessindustrie. atp 61(1–2): 70–81. https://doi.org/10.17560/atp.v61i1-2.2396
46. Shafto M, Conroy M, Doyle R et al. (2010) DRAFT modeling, simulation, information, technology and processing roadmap. https://www.nasa.gov/pdf/501321main_TA11-MSITP-DRAFT-Nov2010-A1.pdf
47. Tao F, Zhang M, Nee AYC (2019) Digital twin driven smart manufacturing. Academic Press
48. Thomas M (1984) Erfahrungswissen versus Planungswissen: Facharbeiterkompertenz und informationstechnologische Kontrolle am Beispiel betrieblicher Instandhaltung. In: Jürgens U, Naschold F (eds) Arbeitspolitik: Materialien zum Zusammenhang von politischer Macht, Kontrolle und betrieblicher Organisation der Arbeit, 1st edn. Westdeutscher Verlag, Opladen, pp 231–251
49. Verband Deutscher Maschinenbauer (2017) Predictive Maintenance: Service der Zukunft—und wo er wirklich steht. https://industrie40.vdma.org/documents/4214230/17409951/1495448864389_VDMA_Predictive_Maintenance_D.pdf/559d1b0e-ea06-4a54-ac52-43bbeb267a7c

50. Weinrauch M (2005) Wissensmanagement im technischen Service: Praxisorientierter Gestaltungsrahmen am Beispiel Industrieller Großanlagen, Gabler. Deutscher Universitätsverlag, Wiesbaden, Wissenschaft. Strategisches Kompetenz-Management
51. Weinzierl S (2019) So nutzt der Instandhalter Datenbrillen und Tablets. https://www.instandhaltung.de/organisation/so-nutzt-der-instandhalter-datenbrillen-und-tablets-111.html. Accessed 05 Feb 2020
52. Windelband L, Dworschak B (2015) Arbeit und Kompetenzen in der Industrie 4.0 Anwendungsszenarien Instandhaltung und Leichtbaurobotik. In: Hirsch-Kreinsen H, Ittermann P, Niehaus J (eds) Digitalisierung industrieller Arbeit, 1st edn. Nomos Verlagsgesellschaft, Baden-Baden, pp 72–87

Martin Eisenmann is a junior researcher at the TU Dortmund University (Social Research Center) and is pursuing his Ph.D. in the DFG-funded research group "Adaption Intelligence of Factories in a Dynamic and Complex Environment" (GRK 2193). Until 2016, he studied social sciences at the Ruhr University Bochum and was a research assistant of the chair of economic and industrial sociology at the TU Dortmund University. His research focuses on work and industrial sociology and includes topics such as digitalization of industrial work, logistics 4.0 and smart maintenance.

Nick Große Nick Große has been research assistant of the Chair of Enterprise Logistics (LFO) at the TU Dortmund University. He is also a member of the research training group "Adaption Intelligence of Factories in a Dynamic and Complex Environment" (GRK 2193). His research focuses on approaches for the joint coordination of production and maintenance in distributed networks.

Towards a Systematic Approach for Chatbot Development in Digital Work Environments

Ingo Börsting and Marc Hesenius

Abstract Chatbots allow to interact with machines in a natural language dialog. Especially in digital work environments, they offer potential for a more intuitive human-computer interaction and remove the need for artificial user interfaces, which users have to learn to operate. However, chatbots typically have to be configured with knowledge about the use case at hand, potential dialog paths, and user interactions, which means to explicitly define user input and chatbot response. The interplay between human and machine defines the perceived user experience. In this chapter, we describe our approach to develop dialogs for chatbot interaction. We use and adapt the Interaction Room Method for this purpose and introduce necessary elements to define required components, especially utterances, intents, and entities.

Keywords Chatbots · Development · Interaction Room

1 Introduction

Over the past decades, Human-Computer Interaction (HCI) has evolved from the use of punchcards containing a set of commands with no interaction at all (except input and output), over text-based command line interfaces, to the modern era of graphical User Interfaces (UIs) following the WIMP[1]-paradigm [22]. But new HCI methods are making their way from research into mainstream computing. New interfaces allowing for a more natural and intuitive way of HCI are becoming available and make their way into everyday applications—users hopefully no longer will have to adapt to UIs,

[1]Windows, Icons, Menus, Pointer—the common interaction paradigm used in most computing systems.

I. Börsting (✉) · M. Hesenius
University of Duisburg-Essen, Schützenbahn 70, 45127 Essen, Germany
e-mail: ingo.borsting@uni-due.de
URL: http://se.wiwi.uni-due.de

M. Hesenius
e-mail: marc.hesenius@uni-due.de

M. Klumpp and C. Ruiner (eds.), *Digital Supply Chains and the Human Factor*, Lecture Notes in Logistics,
https://doi.org/10.1007/978-3-030-58430-6_5

but instead UIs adapt to their users. However, developing such systems with UIs beyond the WIMP-paradigm requires meticulous engineering and consideration of different perspectives, especially with regard to the individual User Experience (UX). Users expect a *good feeling* when using a system—a rather fuzzy term particularly affected by HCI [16], but unpredictable with classic engineering methods [2].

A particularly intriguing HCI component are chatbots: Systems that allow users to communicate in a natural language manner with virtual entities that often present themselves as virtual assistants. They profoundly change the way we interact with machines,[2] because users do not need to adapt to a certain UI by learning its functions, but instead the machine is adapted to a human interaction style (which can be emphasized by mimicking well-known messaging applications, e.g. the use of bubbles for displaying answers). Chatbots are thus interesting for digital work environments to remove the barriers inherently existing in modern computing systems and employ a way of HCI incorporating natural human interactions. Furthermore, recent research hints that cooperating humans and machines achieve better results than *human first* or *machine first* approaches [15], thus a natural matter of interaction is necessary to increase productivity. While speech-based interfaces are an interesting option, common chatbots nowadays are limited to text-based input and output, which will also be the focus of this chapter. Thus, with the ongoing digitalization of industrial branches like production, logistics, and supply chain management, chatbot-like systems are already used for e.g. customer care and first-level support, but offer interesting capabilities for more applications and especially a deeper integration into everyday workflows.

Todays chatbots, however, do not simply *work* for every imaginable use case, instead they have to be specifically designed to serve a precisely defined purpose. Expected dialogs and suitable answers to user queries have to be developed, tested, and deployed and chatbot creation must revolve around them [7]. Developers need to define what input can be expected from users and how the chatbot should react. The input and the various decisions leading to the chatbot's answers can be visualized as a large graph representing the chatbot's capabilities (see Fig. 1 for an example). Within this graph, a dialog is only one of several possible paths depending on how the interaction between user and chatbot progresses. Obviously, the definition of an efficient chatbot users perceive as a proficient and helpful assistant requires domain-specific expertise, therefore domain experts and software developers must work together to create a suitable solution. Developing chatbots thus boils down to classic software development, where agile methods are quite popular nowadays [11] for interdisciplinary teams.

In this chapter, we describe our approach to create chatbots. We build on the Interaction Room [3], a tool for agile software development, and added necessary elements for chatbot development: canvases for the core concepts (intents, dialogs, entities, and integrations) and a set of annotations to introduce perspectives and visualize important concepts. We also describe in detail how our changes should be used in workshops with domain experts and stakeholders. We thus contribute to

[2]We will use the terms *Computer* and *Machine* as synonyms throughout the chapter.

the field of human factors in software development in general and especially to the development of digital work environments in particular by providing a method to develop chatbot interaction in interdisciplinary teams.

We will first introduce some necessary background (chatbots and the Interaction Room) in Sect. 2. We then describe our approach to chatbot development in Sect. 3 and discuss the relation to recent HCI findings in Sect. 4. We conclude the chapter with an outlook for future research options in Sect. 5.

2 Background

We will now introduce some fundamentals necessary to understand our approach to chatbot development. First, we describe chatbots in general and what aspects are required to develop them (Sect. 2.1). We then describe the basics of the method we extended, the *Interaction Room* (Sect. 2.2).

2.1 Chatbots

Chatbots are typically seen as an application of Articial Intelligence (AI).[3] In fact, the idea of a chatbot has accompanied AI development since the very beginning: 1966 Joseph Weizenbaum introduced ELIZA, a program capable of having a dialog-based conversation [23]. ELIZA took a user's input, analyzed its content, and answered the query typically with a question to continue the dialog. Dialogs typically centered around the user's emotional status as ELIZA was initially modeled after a psychotherapist. However, ELIZA merely fooled users to believe it would really understand the dialog: although it reacted with a (mostly) fitting counter-question, its core was a transformation algorithm capable of taking the input and generating a corresponding new question. Nevertheless, ELIZA shows how a dialog between human and machine can be achieved, albeit in a very limited application domain and without a glimpse of real *intelligence*. Over the course of years, similar systems with better performance (i.e. user acceptance) have been developed [19].

Todays chatbots, however, surpass ELIZA's skills and can be actually helpful instead of merely transforming the user's input into an appropriate question to keep the dialog going. Often coming in the form of *virtual assistants*, chatbots are used for e.g. customer support and answering specific questions about products and services. However, chatbots offer more interesting capabilities and also have the potential to enable next generation UIs–see for example the famous *Computer* from Star Trek, which would assist Starfleet personell in an intuitive manner and was also capable to ask e.g. for more input if necessary. While such systems are worth striving for, we

[3]Although this classification is often questionable, because the foundation of most chatbot applications is a classic rule-based system without AI involvement.

will focus on simple text-based chatbots for now, because other systems introduce new technical hurdles, particularly voice processing. Typical chatbots require the definition of three core elements: utterances, intents, and entities.

We refer to the user's input as an *utterance*, thus an utterance is essentially everything the user types into the interface. The utterance is the basic element that triggers all further processing. First, the chatbot system has to tokenize[4] the input and identify keywords it knows. It might also add more context to the keywords, which is often used to boost chatbot performance. Here, AI components come into play, as the conversation itself can be considered a classical rule-based information system, but identifying necessary context often requires pattern matching, which is what AI system do best. In any case, developers must define which utterances to expect, but the definition of suitable utterances may be tricky for several reasons: users might use synonyms, interact in a very individual style, or make typos and grammatical errors.

Having received the user's utterance, the chatbot needs to identify the user's *intent*. An intent is a generalized form of an utterance. In simple terms, many utterances can lead to the same intent, for example, the utterances *Hi, Bot!* and *Hey there!* from the user will be translated into the intent *Greeting*. Within the dialog, the intent is what the chatbot developer is actually looking for and what determines the chatbot's reaction. However, it is as important to have the chatbot react appropriately when the intent cannot be identified.

As mentioned before, to further detail identified intents, the chatbot analyzes the utterance and aims to identify suitable keywords. Among keywords are some with special meaning for the dialog; these are called *entities*. Entities are used to add details to intents and are thus important for the chatbot when deciding for an appropriate answer. When defining the necessary entities, the problem with synonyms becomes especially apparent.

Developers thus need to specifically connect utterances with intents and entities to have chatbots perform their desired task. Current chatbots are highly specialized conversational partners trained to perform certain duties with a set of knowledge actually restricted to this very task—to the best of our knowledge, all-purpose chatbots do not currently exist. Mastering the art of chatbot development thus requires developers to create a system that users perceive as a capable assistant providing useful (and correct) answers, which in turn requires to introduce domain-specific knowledge. As typical in software development, software developers need to work hand in hand with domain experts to create successful solutions. Their main task is to define the dialog flow between user and chatbot. Figure 1 shows an example dialog flow from a rather simple chatbot, whose main task is to answer support queries for an enterprise resource planning system. When encountering errors in the system, the chatbot serves as a first-level support that users can ask for potential solutions. The example shows how complex the underlying dialog graph can become even for such a simple chatbot. Figure 2 shows a configuration detail, where the chatbot asks the user for the specific error message, analyzes the corresponding utterance to

[4]Tokenization is the process of identifying meaningful entities with a text; in this case, words.

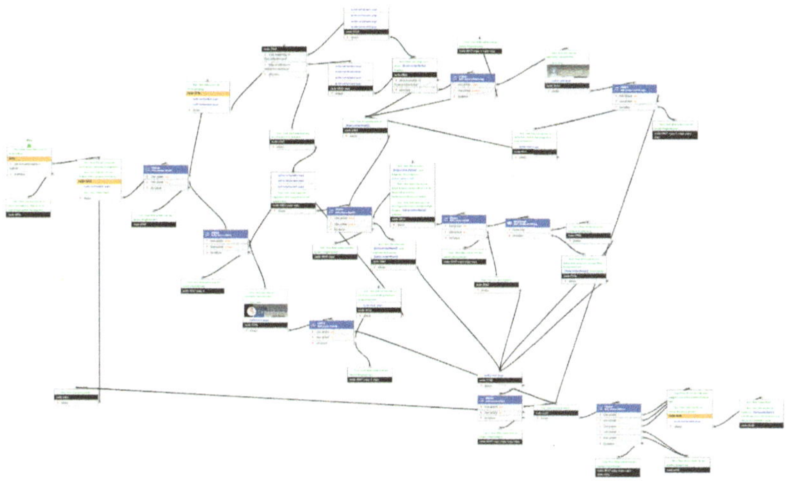

Fig. 1 Overview of a chatbot dialog system [modeled with the open source chatbot framework *BotPress* (https://botpress.com)]

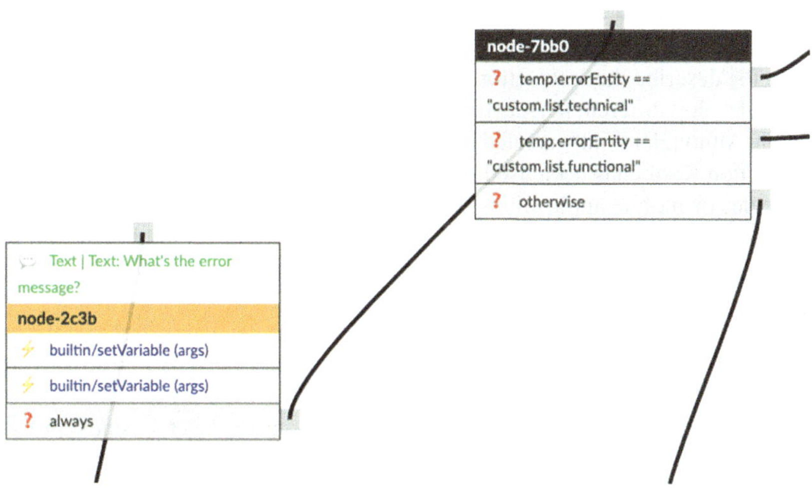

Fig. 2 Details of a chatbot dialog graph

identify whether the user is having a technical (e.g. *the server is not responding*) or functional (e.g. some wrong input) error message, and acts accordingly.

While the technology to create chatbots is widely available, chatbot development projects often fail to deploy successful, i.e. widely accepted, chatbots for various reasons. Using chatbots means to employ a different style of HCI, which naturally comes with different requirements and expections from users [4]. Development teams need to focus on an actual conversation [7] instead of the simple command-and-react style of common graphical UIs following the WIMP-paradigm (see also Section 4 for further HCI related aspects and how we included them in our approach). Another problem is the mismatch between user expectations and actual chatbot performance [17], which in turn requires developers to carefully manage how conversations between user and chatbot progresses and to explicitly communicate the chatbot's limitations. Managing user expectations requires all stakeholders to develop a common vision of what the desired application is capable of—a core principle in agile software development methods.

2.2 Interaction Room

Developing software requires communication between domain experts, software developers, and other experts in their respective field (e.g. considering regulation or special types of UI design). Sharing a common vision is crucial for successful project teams, especially with regard to digital work environments that not only require acceptance of single applications, but of overall digital workflows. Teams thus have to maintain the *big picture* and review the desired system from several perspectives. Modern software development employs *agile methods*[5] to structure the work in software teams, but these methods still require tools for teams to work with. Book et al. [3] describe the Interaction Room as a tool software development teams can use to develop different perspectives, distribute knowledge, and maintain a common vision. Although initially created to gather requirements in a structured approach, the Interaction Room has been used for various purposes, e.g. to develop digitalization strategies or mobile applications [3]. Also, a digital version (the Augmented Interaction Room (AugIR) [13]) has been developed. It enhances the Interaction Room with the capabilities of digital whiteboards and is used to e.g. connect UI sketches and business processes [14] or define modern gesture-based UIs [9, 10].

Book et al. [3] describe the Interaction Room as a physical room for the team (including all stakeholders) to gather and exchange information, discuss project matters, and especially visualize the various aspects involved. A common Interaction Room is outfitted with whiteboards that the team uses to sketch business processes, digital and virtual objects, interactions between and with surrounding systems, etc. Notations for visualization are kept simple with boxes, lines, and free syntax to lower any communication barriers introduced by more formal notations—in fact, formal notations have been reported to be disruptive to the general workflow in early software-development[6] [20]. Thus, the Interaction Room emphasizes the interaction

[5] Agility is more of a mindset than a concrete method. In simple terms, using agile methods should enable teams to develop and deploy software in smaller chunks, which should lead to fewer errors and faster integration of new requirements. The *Manifesto for Agile Software Development* summarizes the core ideas [1].

[6] Which by no means implies that formal notations are useless, but that they should be used when appropriate, e.g. to document finalized models.

between stakeholders to distribute information between team members. While the method itself is not strictly defined, some principles are worthwhile mentioning. Most prominently, the Interaction Room is not a *free discussion*, but employs two special roles, namely the *Method Coach* and the *Domain Coach*. Method coaches are the workshop chairs and ensure that all stakeholders receive appropriate time to introduce their perspective and discuss relevant aspects. The domain coaches' main task is to keep the appropriate level of abstraction, i.e. to ensure that stakeholders do not deep-dive into unnecessary details, which in turn requires them to have knowledge of the respective business domain (hence the name).

To structure the different stakeholder perspectives in the Interaction Room, Book et al. [3] propose to use *Canvases* and *Annotations*. A canvas is a visualization on one of the whiteboards. Depending on the project's goal and the Interaction Room incarnation, different canvases can be used to e.g. identify necessary partners and their interfaces (Partner Canvas), define surrounding software systems (Integration Canvas), or describe business processes (Process Canvas). Information sketched on the canvases is typically annotated with a special set of symbols afterwards. Stakeholders can use annotations to express their perspective, opinions, and concerns. According to Book et al. [3], annotations may be *Value Drivers*, e.g. parts of the system that will influence the business or user value, or *Effort Drivers*, e.g. boundary conditions that must be met (regulatory requirements would be a typical effort driver). Another important aspect is that stakeholders can annotate *Uncertainty*, which emphasizes critical parts, e.g. when parts of processes cannot be defined explicitly.

3 Using an Interaction Room to Develop Chatbots

Within our approach to develop chatbots systematically, we adapt basic elements of the Interaction Room (see Sect. 2.2) and extend them with specific artifacts necessary to discuss relevant topics, namely the aforementioned utterances, intents, and entities (see Sect. 2.1). Furthermore, we argue that potential sources of information, i.e. surrounding systems, must be discussed for proper integration into digital work environments. Our extensions include different canvases, each of which defines a specification subset, and a set of annotations to reflect artifacts, foster a common understanding of the chatbot's purpose, and highlight potential problems and benefits.

We use an example chatbot to illustrate the use of our approach: it answers support requests and classifies them as functional or technical inquiries to provide targeted support. The chatbot should be able to determine a support staff member to transfer the conversation to if the chatbot's assistance is insufficient. Core artifacts of the concept are explained on the basis of this example in the following. We will first introduce the canvases for the main artifacts: *Intent Canvas* (Sect. 3.1), *Dialog Canvas* (Sect. 3.2), and *Entity Canvas* (Sect. 3.3). We then describe how we adapt the Interaction Room's *Integration Canvas* (Sect. 3.4) and which annotations we propose (Sect. 3.5).

3.1 Intent Canvas

Chatbots classify user intents to react to them appropriately (see Sect. 2.1). In order to tailor a chatbot's reactions to an intent, each intent requires an individual dialog tree reflecting specific user needs for this intention and the conversation leading to its satisfaction. Within our concept, all intents that should be represented by the chatbot are assembled on the *Intent Canvas*, which is inspired by the Interaction Room's *Feature Canvas* [3]. Intents are typically assembled before the actual workshop session in consultation with selected stakeholders (e.g. product owners), however, discussing the Intent Canvas with participating stakeholders is a good kick-off for a workshop to discuss Intent priorities (which defines the workshop schedule) and clarify potential conflicts. Following the principles of agile software development, the *Intent Canvas* is highly dynamic and intents can be added, changed, or dismissed during the various workshop sessions. Within sessions, individual intents are selected as a starting point for the actual definition on the other canvases. In our example, the support chatbot, possible intents could be *Error Support*, *User Manual*, or a few more. Figure 3 shows an example.

3.2 Dialog Canvas

With the *Dialog Canvas*, we model a chatbot's core feature: the dialog itself. For every previously defined intent, the team creates a corresponding *Dialog Canvas*. Dialogs are represented like activity diagrams, where each part in the conversation is marked as chatbot (B) user (U) activity. The exemplary dialog flow shown in Fig. 4 depicts a dialog in which a user reports an error after being greeted by the chatbot.

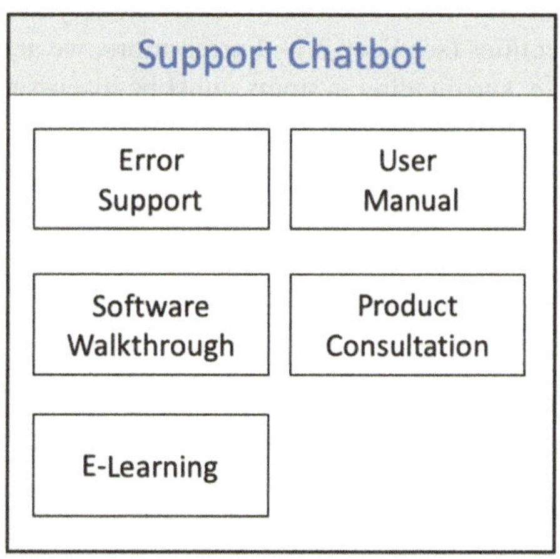

Fig. 3 A potential *Intent Canvas* for a support chatbot

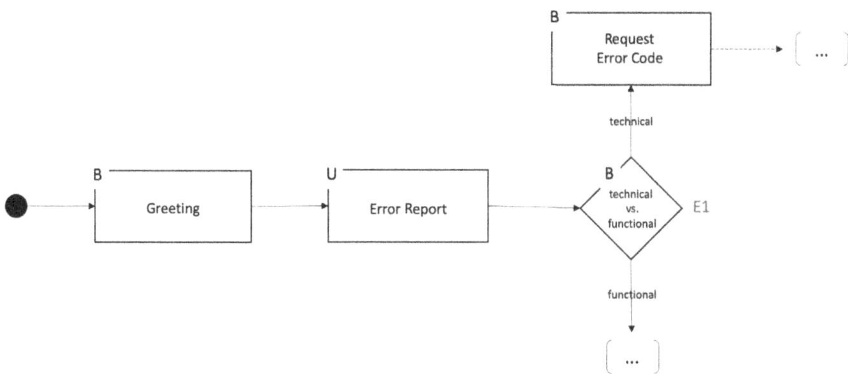

Fig. 4 Extract of an exemplary dialog tree on the *Dialog Canvas*

Thus, the first activity *Greeting* is marked as a chatbot action and the action *Error Report* as a user action.

In chatbots, decision nodes are crucial and determine the further course of a dialog. Decisions are either based on user or chatbot actions. The dialog flow in Fig. 4 contains a decision node where the users' error report is analyzed to decide whether the error is of technical or functional nature. This decision is marked as a chatbot action, thus the chatbots internal logic differentiates between both. Such chatbot decisions are usually made on the basis of the information acquired during the conversation (*Entities*, see Sect. 2.1), thus the development team must have a special focus on it. On the dialog canvas, chatbot decisions based on entities are marked with a unique entity identifier (in the example *E1*, see Fig. 4). Assigning entities to decision nodes simplifies the collection of required data and enables the structured recording and elaboration of entities on the *Entity Canvas* (see Sect. 3.3).

In line with Interaction Room principles, the workshop coaches are in charge of guiding the stakeholders during dialog modeling. However, the participants should define the content. As mentioned above, our approach particularly distinguishes between user and chatbot activities and considers decision nodes and their foundation for decision-making.

3.3 *Entity Canvas*

Decisions are crucial for a chatbot dialog and are usually based on supplied data during the conversation. We use the *Entity Canvas* to highlight the significance of decisions made in dialogs and to facilitate the collection of data required for this purpose.

The dialog in Fig. 4 illustrates a chatbot that classifies a reported error as either of technical or functional source. Entities relevant for this decision are marked within

E1 – Error Classification	
technical	Connection Timeout Server Error [*Internal Server Error, Bad Gateway*] Unauthorized [*Authentication Failed, Network Authentication Required*] Forbidden
functional	Wrong VAT Code [*Incorrect VAT Code, Wrong Value Added Tax Code*] VAT Invalid Tax Entries Are Incomplete

Fig. 5 Extract of an exemplary entity on the *Entity Canvas*

the dialog and listed on the *Entity Canvas* in parallel, considering the unique entity identifier defined within the dialog tree (here *E1*, see Fig. 4). As shown in Fig. 5, the entities for this decision are divided into two categories: technical and functional errors. Stakeholders then discuss and assemble various forms of typical error messages from these entity categories. Synonyms can be specified for each entity attribute, which are appended to the corresponding value.

It is important to note that it is not required to create a complete entity list with all imaginable values. The assembly of entity values serves the shared overview and common understanding of representative characteristics of individual entity categories. Entity values can be added after the chatbot specification or imported from surrounding systems (see Sect. 3.4).

3.4 *Integration Canvas*

Depending on the use case at hand, chatbots can be integrated into the existing application landscape and use data that is fed from other systems via interfaces in order to make decisions or display targeted information. At the same time, information from chatbot dialogs can be transferred to surrounding systems, e.g. to document the conversation and thus improve the chatbot specification through analyzing past conversations. Within our approach, the *Integration Canvas* from the original Interaction Room serves the documentation of surrounding systems by visualizing information flows and types.

As shown in Fig. 6, the chatbot is centered on the *Integration Canvas*, while all external systems are arranged around the center. Here, applications that obtain information from or feed information into the chatbot are visualized. In the described example, it is conceivable that experts' contact information is taken from the *Active Directory* and displayed when a dialog endpoint is reached and the conversation is to be transferred to a human entity. On the *Integration Canvas*, information flows are visualized by arrows that indicate the direction of flow. Since the information from the *Active Directory* flows into the conversation, the arrow is directed from the surrounding system into the chatbot (see Fig. 6). In this example, it is further

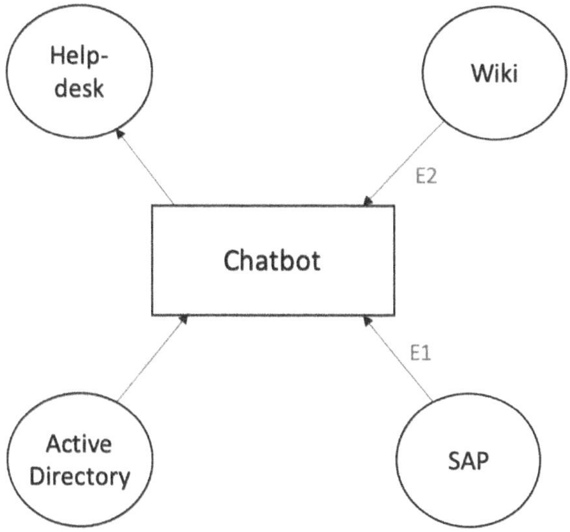

Fig. 6 Extract of an exemplary *Integration Canvas*

conceivable that error reports should automatically be forwarded to a *Helpdesk*, if the error could not be resolved by the chatbot itself. Thus, an information flow from the chatbot into the surrounding *Helpdesk* is documented. As explained in Sect. 3.3, entities are essential for chatbot dialogs. Surrounding systems can be used to feed entity values into the chatbot system. On the *Integration Canvas*, entity identifiers from the *Dialog Canvas* are used to reference entities fed from surrounding systems to the dialog tree. The entity objects listed in Fig. 5 are therefore not manually fed into the system but rather imported from the *SAP* system listed on the *Integration Canvas*.

Through the specification of surrounding systems on the *Integration Canvas*, a common understanding of essential interfaces and information flows between applications is generated. Resulting requirements for the chatbot and the application landscape become evident.

3.5 Annotations

We introduce new annotations to reflect and evaluate the chatbot specification and to document the various perspectives from the stakeholders. To meet the requirements of a systematic chatbot specification, annotations for chatbot-specific aspects are needed. These annotations should reveal and clarify requirements regarding complexity, entities, endpoints, and content of the desired chatbot.

Table 1 Additional annotation for systematic chatbot specification

	Synonyms	Marks dialog activities that require a high amount of synonyms, e.g. to capture the end user intent or as a foundation for chatbot decisions
	External resource	Marks dialog activities that require a connection to external resources, e.g. to integrate multimedia content or feed entities/synonyms
	Human interaction	Marks dialog endpoints in which the conversation is to be transferred to a human entity
	Multiple choice	Marks dialog activities where users should choose from a set of defined response options
	Free text	Marks dialog activities in which a free text input should be integrated

Table 1 lists and explains the additional annotations used in our approach. These annotations can be used on all aforementioned canvases. In line with the Interaction Room Method, the discussion between stakeholders about the annotations is documented and requirements are derived accordingly.

4 Relation to HCI Design for Chatbots

When specifying chatbots, it is essential to be aware of how users interact in conversations [7]. We created our approach to systematic chatbot development taking current findings in the field of HCI into account. In the following, various findings of chatbot-related research are described, and their consideration in our approach is discussed.

Toxtli et al. [21] found that users typically converse with a chatbot in a human-like manner. Here, users often use courtesies such as *please* and *thank you*, although emotionless machines will obviously have no *bad feelings* if omitted. Thus, when developing chatbot dialogs, human-like attributes must be considered in order to reflect users' communication patterns. In our approach, dialog artifacts that have a special need for human-like behavior can be marked by appropriate annotations: A large number of synonyms could be used to create a broader vocabulary of the chatbot and thus lead to a more natural conversation style. We provide a corresponding annotation for *Synonyms*. In addition, free text inputs can be marked within the dialog which should allow a more free and thus natural conversation style. This is provided by the annotation *Free text*. In the development of chatbots, it is necessary to distinguish between a free natural conversation style and the potential for errors caused by a user-driven conversation and a thus more complex text analysis. Findings in this regard will follow in the course of this section.

Kim et al. [12] highlight the fundamental influence of dialog modeling on the satisfaction of end users. A chatbot that can adapt to users' emotional states can strongly influence the perceived conversation quality. This concerns the conversation tone and style, since users perceive computers as social actors and thus sense a high social presence in a chatbot conversation. According to the authors, a chatbot should be tailored to a specific use case in order to efficiently classify and mirror the user's emotional state. In our approach, the individual consideration of intents and the specific modeling of dialogs per intent allows to specify the conversation style as well as the social reflection individually to the use case pursued in the intent.

The potential of user-driven conversations must be weighted against the complexity of error-prone text recognition. Cerezo et al. [5] state that user-driven conversations often fail because users expect the chatbot to be able to react to a wide variation of input, which unfortunately most chatbots cannot provide. Fiore et al. [6] add that the main cause for negative UX in a chatbot conversation is misclassifications of user input. In simple terms, the chatbot does not understand the user. A typical cause is the use of too many open questions, whose technical analysis (i.e. identification of intent and entities) is more error-prone and thus increases the chance for user frustration. Additionally, in use cases where the chatbot serves as first level support (e.g. customer support), users might already be frustrated when starting the conversation. Misunderstandings and incorrect answers could further strengthen already existing negative emotions. Considering these findings, conversations should be transformed into a chatbot-driven form, i.e. the chatbot must take control of the dialogue and guide the user into giving expected answers. As Fiore et al. [6] state, participants appreciate simplicity and pro-active guidance as well as immediate feedback in order to accomplish tasks. We considered these findings by enabling stakeholders to annotate dialog artifacts that enable a chatbot-led conversation. For example, the annotation *Multiple Choice* can be used to mark dialog items where the conversation should be restricted to a predefined set of options to allow a chatbot-led dialog. Furthermore, the developed dialog flow should indicate where potential points of frustration, especially open questions and free text input, can be found. Dialog artifacts for which free text user input is indispensable can be marked using the corresponding annotation (*Free text*, see Section 3.5) and can additionally be labeled with the annotation *Synonyms* in order to highlight elements where a broad variation of user input is expected.

Følstad et al. [8] further state that a chatbot should clearly communicate what it is capable of in order to avoid frustration. Here, the transparent communication of skills helped to limit possible frustration through unrecognized intentions. Thus, we provide the annotation *Human Interaction* to mark endpoints of a chatbot's ability, which should be communicated transparently. Stakeholders should annotate dialog endpoints that need the conversation to be transferred to a human entity. By annotating these endpoints, it becomes evident what the chatbot is capable of and what it is not.

This serves to communicate the chatbot's capabilities within the conversation to the user and allows to reduce potentially frustrating experiences.

5 Conclusion

We described our approach to systematic chatbot development using the Interaction Room [3] Method. For this purpose, we extended the Interaction Room with necessary details to gather information about chatbot dialog paths and user interaction. Domain experts and software developers can use our approach to model the behavior of chatbots and to understand how utterances, intents, and entities must be defined for the chatbot to properly do its job. The dynamic nature of our approach allows us to incorporate future findings from HCI and chatbot research and to introduce necessary extensions, e.g. additional annotations.

With the market for chatbot applications increasing in the next years [18], the experience when interacting with chatbots must be tailored to the desired use case as well as the users' expectations. To avoid frustrating experiences and dissatisfied customers, development teams must focus on dialog creation and identify potential pitfalls early. Our method shows how to systematically develop chatbots and thus offers interesting potential for future applications and the creation of satisfying HCI, especially in digital work environments.

For future work, several interesting tasks remain. We will validate our approach in several chatbot development projects and make necessary adjustments according to the results obtained. Furthermore, as the Interaction Room has a digital sibling (the AugIR [13]), an interesting option would be to digitize our approach as well and derive automatic configuration files for chatbots.

References

1. Beck K, Beedle M, van Bennekum A, Cockburn A, Cunningham W, Fowler M, Grenning J, Highsmith J, Hunt A, Jeffries R, Kern J, Marick B, Martin RC, Mellor S, Schwaber K, Sutherland J, Thomas D (2001) Manifesto for agile software development. http://www.agilemanifesto.org/. Accessed 20 April 2020
2. Blandford A (2013) Engineering works: what is (and is not) "Engineering" for Interactive Computer Systems? In: Proceedings of the 5th ACM SIGCHI Symposium on Engineering interactive computing systems. In: EICS'13. ACM Press, New York. https://doi.org/10.1145/2494603.2480342
3. Book M, Gruhn V, Striemer R (2016) Tamed agility: pragmatic contracting and collaboration in agile software projects. Springer International Publishing, Berlin
4. Brandtzaeg PB, Følstad A (2018) Chatbots: changing user needs and motivations. Interactions 25(5):38–43. https://doi.org/10.1145/3236669
5. Cerezo J, Kubelka J, Robbes R, Bergel A (2019) Building an expert recommender chatbot. In: Proceedings of the 1st international workshop on bots in software engineering. In: BotSE'19. IEEE Press, Montreal, May 2019, pp 59–63. https://doi.org/10.1109/BotSE.2019.00022

6. Fiore D, Baldauf M, Thiel C (2019) Forgot your password again?: acceptance and user experience of a chatbot for in-company IT support. In: Proceedings of the 18th international conference on mobile and ubiquitous multimedia (MUM' 19), Association for Computing Machinery, Pisa, Italy, Nov 2019, , pp. 1–11. https://doi.org/10.1145/3365610.3365617
7. Følstad A, Brandtzæg PB (2017) Chatbots and the new world of HCI. Interactions 24(4):38–42. https://doi.org/10.1145/3085558
8. Følstad A, Nordheim CB, Bjørkli CA (2018) What makes users trust a chatbot for customer service? An exploratory interview study. In: Bodrunova SS (ed) Internet science. Lecture notes in computer science. Springer International Publishing, Cham, pp 194–208. https://doi.org/10.1007/978-3-030-01437-7_16
9. Hesenius M, Gruhn V (2019) Gesturecards—a hybrid gesture notation. In: Proceedings of the ACM on human-computer interaction 3:22:1–22:35. https://doi.org/10.1145/3331164
10. Hesenius M, Kleffmann M, Gruhn V (2016) Sketching gesture-based applications in a collaborative working environment with wall-sized displays. In: 2016 International conference on collaboration technologies and systems (CTS). IEEE Conference Publications, pp 327–336. https://doi.org/10.1109/cts.2016.0066
11. Hoda R, Salleh N, Grundy J, Tee HM (2017) Systematic literature reviews in agile software development: A tertiary study. Inform Software Technol 85:60–70. https://doi.org/10.1016/j.infsof.2017.01.007
12. Kim S, Lee J, Gweon G (2019) Comparing data from chatbot and web surveys: effects of platform and conversational style on survey response quality. In: Proceedings of the 2019 CHI conference on human factors in computing systems. In: CHI '19, Association for Computing Machinery, Glasgow, Scotland UK, May 2019, pp 1–12. https://doi.org/10.1145/3290605.3300316
13. Kleffmann M (2014) Augir: the conceptual design and evaluation of an augmented interaction room. In: Proceedings of the 29th ACM/IEEE international conference on automated software engineering. In: ASE '14, Association for Computing Machinery, New York, NY, USA, pp 883–886. https://doi.org/10.1145/2642937.2653467
14. Kleffmann M, Hesenius M, Gruhn V (2015) Connecting UI and business processes in a collaborative sketching environment. In: Proceedings of the 7th ACM SIGCHI symposium on engineering interactive computing systems. In: EICS' 15. ACM, New York, NY, USA, pp 200–205. https://doi.org/10.1145/2774225.2775076
15. Klumpp M, Hesenius M, Meyer O, Ruiner C, Gruhn V (2019) Production logistics and human-computer interaction–state-of-the-art, challenges and requirements for the future. Int J Adv Manuf Technol 105(9):3691–3709. https://doi.org/10.1007/s00170-019-03785-0
16. Law ELC, Roto V, Hassenzahl M, Vermeeren APOS, Kort J (2009) Understanding, scoping and defining user experience: a survey approach. In: Proceedings of the SIGCHI conference on human factors in computing systems. In: CHI '09, ACM, New York, NY, USA, pp 719–728. https://doi.org/10.1145/1518701.1518813
17. Luger E, Sellen A (2016) "Like having a really bad PA": the gulf between user expectation and experience of conversational agents. In: Proceedings of the 2016 CHI conference on human factors in computing systems. In: CHI '16, Association for Computing Machinery, New York, NY, USA, pp 5286–5297. https://doi.org/10.1145/2858036.2858288
18. Nguyen MH (2020) The latest market research, trends, and landscape in the growing AI chatbot industry. https://www.businessinsider.com/chatbot-market-stats-trends. Accessed 20 April 2020
19. Shah H, Warwick K, Vallverdú J, Wu D (2016) Can machines talk? Comparison of Eliza with modern dialogue systems. Comput Hum Behav 58:278–295. https://doi.org/10.1016/j.chb.2016.01.004
20. Shipman FM, Marshall CC (1999) Formality considered harmful: Experiences, emerging themes, and directions on the use of formal representations in interactive systems. Comput Support Cooperative Work 8(4):333–352
21. Toxtli C, Monroy-Hernández A, Cranshaw J (2018) Understanding chatbot-mediated task management. In: Proceedings of the 2018 CHI conference on human factors in computing systems.

In: CHI'18. Association for Computing Machinery, Montreal, QC, Canada, pp 1–6, April 2018. https://doi.org/10.1145/3173574.3173632
22. van Dam A (1997) Post-WIMP user interfaces. Commun ACM 40(2):63–67. https://doi.org/10.1145/253671.253708
23. Weizenbaum J (1966) Eliza—a computer program for the study of natural language communication between man and machine. Commun ACM 9(1):36–45. https://doi.org/10.1145/365153.365168

Ingo Börsting is a Ph.D. student at the University of Duisburg-Essen in Germany. He studied Information Management at the University of Applied Sciences in Hanover and received his master's degree in applied cognitive and media science at the University of Duisburg-Essen, where he currently holds a position as a research assistant at the chair of Software Engineering. His research focuses on new aspects of software engineering for emerging and novel technologies, including cyber-physical and chatbot systems as well as mixed- and virtual reality software.

Marc Hesenius is a postdoctoral researcher in the field of software engineering at the University of Duisburg-Essen. His research interests focus on methods and tools to engineer data-driven applications and the technical aspects of modern Human-Computer Interaction.

Utilization

Digital Supply Chains and New Work—Friends or Foes?

Caroline Ruiner and Christina E. Debbing

Abstract New technologies are introduced in supply chains and thus show the integration of innovations. Usually, these are driven by the management and implemented top-down, workers come into play only to a marginal point. However, work processes often change fundamentally and lead to new work experiences for the workers. Thus, the integration of workers and the consideration of the social system that frames the innovation process is central to the overall success of an innovation. Against this background, it is the aim of this chapter to analyze the worker's attitude toward new technologies in digital supply chains and focuses on the antecedents for a gainful collaboration. We refer to a case study consisting of qualitative interviews with blue-collar workers in a production organization involved in digitalization processes and include observations made on the shop floor. We discuss the findings referring to the diffusion of innovations theory and emphasize the relevance of the worker's participation in designing digital innovations.

1 Introduction

Digital supply chains have various forms and manifestations, such as transport automation, automated storage and picking, technical work support, web/app-based self-organization and AI management [6, 12, 14, 19]. In each form and manifestation, human workers are still involved in processing work tasks and collaborate with digital devices [8]. So far, studies mainly discuss the substitutability of workers, with other investigations emphasizing that the human factor is critical for digitalization processes in organizations [1, 3, 5]. This article analyzes the workers' attitudes toward new technologies in digital supply chains and focuses on the antecedents for a gainful collaboration that they are friends instead of foes. More concretely, we address the question of how to promote workers' acceptance of digital innovations in organizations and how to develop their empowerment. For this, we investigate the adoption

C. Ruiner · C. E. Debbing (✉)
University of Hohenheim, Stuttgart, Germany
e-mail: christina.debbing@uni-hohenheim.de

M. Klumpp and C. Ruiner (eds.), *Digital Supply Chains and the Human Factor*, Lecture Notes in Logistics,
https://doi.org/10.1007/978-3-030-58430-6_6

of innovations by workers and refer in particular to seminal work on the diffusion of innovations [15]. Moreover, we further develop Rogers' theory by highlighting the relevance of the worker's participation in designing digital innovations.

The contributions of this chapter include the transfer of Rogers' [15] theory to the discussion on digitalization, the presentation of empirical findings regarding digitalization processes in the organization, and the identification of antecedents for a gainful collaboration of people working with digital devices.

Our discussion has the following structure. First, we introduce our conceptual framework regarding new work, changing organizations, and the relevance of empowerment, and we present Rogers' [15] Theory on the Diffusion of Innovations (DOI). To answer the research question, we refer to a qualitative study in a production organization, in which a new technology was recently introduced. Then, we discuss the findings with respect to our conceptual framework and our contribution to existing research. We end with suggestions for avenues for further research.

2 Conceptual Framework

Digitalization leads to profound changes in the world of work [4, 13]. In this context, the concept of *new work* addresses the changing values of work and relationships between workers and organizations [7]. So far, literature on new work emphasizes the relevance of values such as individualization, subjectivity, self-determination, and autonomy [2, 11]. From an organizational perspective, this emphasis leads to decentralization, codetermination, interconnectivity, and required participation. A related debate on changing organizations [7, 9] focuses on democratization, de-hierarchization, and stronger self-regulation, with the possibility of developing one's own interests. Organizational changes are necessary to strengthen the participation of workers, as their involvement is critical for the implementation of innovations and productivity [16]. While new technologies might support participation, workers need to be empowered to engage in technology's use.

Empowerment refers to strategies and measures to increase the degree of autonomy and self-determination of workers to represent their interests autonomously, self-responsibly, and self-determinedly. As Fig. 1 shows, Spreitzer [18] assumes that empowerment consists of four dimensions: meaning, competence, self-determination, and impact. *Meaning* is the value of a work goal or purpose and is related to the worker's values. *Competence*, or self-efficacy, is an individual's belief in his or her ability to perform activities with skill. *Self-determination* is an individual's sense of having choice in initiating and regulating actions, and *impact* is the degree to which an individual can influence strategic, administrative, and operational outcomes. Yukl and Becker [21] argue that access to information, funds, materials, and facilities needed to carry out effective work increases the worker's empowerment. Access could be especially promoted through socio-technical systems with flexible technologies.

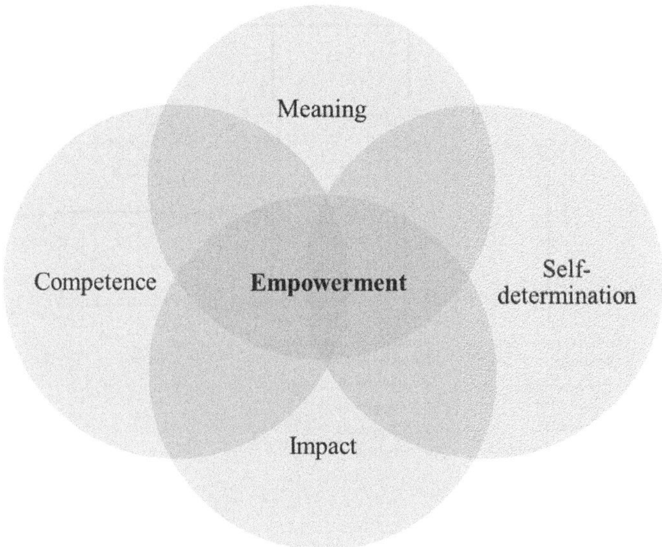

Fig. 1 Empowerment dimensions (own source following [18])

In connecting the changing values of work and highlighting the meaning of workers' empowerment, the DOI theory is applicable to the consideration of workers as crucial success factors in digitalization processes. Within the *innovation decision process* as part of the DOI theory, workers' perspective and their decisions concerning innovations are highlighted. The DOI theory describes "the process by which an innovation is communicated through certain channels over time among the members of a social system" [15, p. 5], from the first time the unit (individual or organization) knows about the innovation until it decides to accept or reject. Throughout the five stages of this process, the DOI theory can be used to analyze motivations, perspectives, and reasoning of individuals and to highlight changes that may occur in work environments where digital innovations are implemented (Fig. 2).

The first stage, *knowledge*, consists of awareness knowledge, how-to knowledge, and principles knowledge [15]. Awareness knowledge is the familiarity with an innovation and the reason for individuals to be motivated to learn more about the innovation. How-to knowledge is the information that individuals need to use the innovation correctly. If the innovation is highly complex and therefore a lot of how-to knowledge is needed, the likelihood of acceptance decreases. Lastly, principles knowledge indicates that the individual understands the main concepts and terms of the innovation and is able to comprehend the innovation from its basis.

The second stage, *persuasion*, refers to individuals forming an attitude and developing conviction considering the attributes of the innovation [15]. The *decision* stage is the state in which the individual decides to accept or reject the innovation. If the innovation is accepted, the individual enters the fourth stage, *implementation*, in which the individual begins using the innovation. In most cases, the decision stage

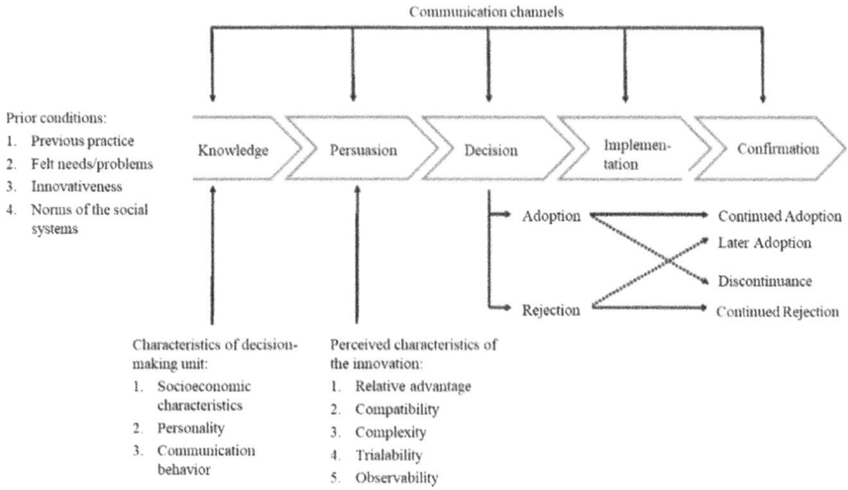

Fig. 2 Innovation decision process [15]

marks the end of the process. However, in cases where the individual experiences dissonance or a form of imbalance, the process extends to a fifth stage of *confirmation,* where the individual seeks information from peers and other information channels to confirm the decision. Thus, peers sharing their assessments within interpersonal channels are of importance: "[m]ost individuals evaluate an innovation not on the basis of scientific research by experts but through the subjective evaluations of near peers who have adopted the innovation" [15].

Through the DOI theory, Rogers [15] provides important indications on how to design innovation implementation processes to promote workers' acceptance. He does so by embedding acceptance in the innovation decision process and thereby highlighting acceptance as part of the implementation process. He also underlines the influences of different stages and stresses that the whole process is complex and socially characterized. Adding the worker's participation into the process of implementing innovations in the context of digitalization allows experiential knowledge to be used for individual process solutions. Additionally, research underlines that the inclusion of workers connects directly to their acceptance: "Evidence indicates that those who have been involved in decisions related to the change may react differently, by being more supportive of the change, than those who were not involved" [17]. Therefore, crucial forces or enablers include the worker's empowerment and opportunities to participate in the implementation process, which will be focused in our study presented in the following.

3 Data and Methods

To answer the question of how workers' acceptance of innovations in organizations can be promoted, we conducted a case study [20] consisting of qualitative interviews with blue-collar workers in a production organization and include observations made on the shop floor. The data were collected through face-to-face semi-structured interviews with five workers who were directly affected by the digitalization innovation. The interviewees were selected because of their different ways of interacting with the innovation.

First, we interviewed two workers who had worked mostly autonomously but with the support of the management to create and implement the innovation. Second, we interviewed two tugger-train drivers because their work changed most owing to changing software and procedures. Last, we interviewed a team leader who functions as the relevant person in case of questions and queries concerning the innovation. Table 1 provides an overview of the sample.

The interviews addressed the workers' focuses, structures of relevance, expertise, and reasoning. The interviews were recorded, transcribed, and analyzed using qualitative content analysis [10] employing the deductive application of codes from the DOI theory.

The production company in this case study changed its intralogistics profoundly. Under the old system, instructions concerning the transportation of constructed parts were shouted out to the drivers and these drivers needed to remember dozens of different transport orders every day. This system made the production hall noisy and quite messy, and left everyone involved unsure of the location of parts and the completion of tasks and orders. To address this disorder, two blue-collar workers took the opportunity to develop a new system that changed the company's tugger-train arrangement. They programmed and installed the necessary software and hardware, then recommended using scanners and programmed an in-house app to optimize procedures, organize tasks, and visualize responsibilities. As one result, the processes that were previously called out in direct communication are now transparent and

Table 1 Sample characteristics

No.	Age	Level of education	Occupation	Leadership responsibility
1	40	Apprenticeship	Equipment engineer (Programming and set up)	None
2	35	Apprenticeship	Equipment engineer (Processing, programming and set up)	None
3	45	Apprenticeship	Team leader (Production) and laser work	Yes
4	45	Apprenticeship	Reserve driver and production worker	None
5	60	Apprenticeship	Tugger train driver	None

traceable for all blue- and white-collar workers of the company. The dealing with this innovative digitalization process and its acceptance is key to our study.

4 Empirical Findings

The data analysis shows multi-layered changes in workers' perception of empowerment owing to the implementation of the innovation and its impact on their work and the participatory design of the implementation itself. To focus on this changing perception of empowerment, we first present the general innovation decision process and then describe the connection between this process and the participatory design of the implementation.

4.1 The Innovation Decision Process in Practice

With respect to knowledge as the first stage of the innovation decision process, the workers emphasized that they had little or no awareness of the innovation and became aware of the innovation by accident or through informal information (4: 24–27). The workers stated that they did have how-to knowledge and principle knowledge about the innovation, for example how barcodes work (3: 11–12) or why scanning of items is important (5: 37:50).

Interviewees readily entered the second stage, persuasion, as they could easily name relative advantages of the innovation such as work relief, improved process workflows, higher transparency and traceability, and less stress. In particular, the workers expressed work relief in terms of a general easing of their work (3: 131–132), clear structuring (4: 168–179), and less concern about the location of produced parts (3: 67–69). When talking about improved process workflows, they highlighted that they liked the new way of coordinating material flow (1: 12–19), the reduction or shortening of standing times (1: 12–19), and greater planning capabilities for workers in general (4: 51–57). The transparency and traceability the innovation provides is exceptional, and the workers stated that they are now able to locate produced parts within the company (2: 117–120), and that after scanning their produced parts they feel relief as they pass on responsibility (5: 53–56) or assign responsibility to their co-workers (2: 103–105). The documentation of the innovation provides them with security or peace of mind concerning liability issues relating to other workers or between workers and their supervisors (2: 91–98). Lastly, the workers feel less stressed because time management is possible (5: 331–332), they experience fewer peak times because of improved smoothness or regularity within their work (4: 168–179), they have a new feeling of control over their work (5: 402–408), and they have a new perspective on their work (4: 51–57).

Concerning the compatibility of the innovation with the social system, changes in collaboration and communication can be highlighted. Interviewees noted that before the innovation was implemented, they worked mostly independently and every worker was able to choose and run processes as he wished, even if the processes were inefficient. With the new work organization, the workers find themselves in a more collaborative working environment, where they scan parts that they produce, finalize production steps, and subordinate their work goals to the overall production goals (2: 290–305). Further, communication has changed from a high percentage of direct communication to communication through the innovation. The workers noted the importance of direct communication (3: 180–186), but mostly stressed that previously direct communication constituted a stressor in their everyday work life (1: 190–201). Lastly, in terms of the stage of persuasion, the observability and the complexity of the innovation influenced the workers' decision process. For the workers, scanning their work and using the connected app on tablets, smartphones, and computers (3: 43–48) gives them the opportunity for transparency and self-control at work. They unanimously described the complexity of the innovation as high at the implementation stage but strongly decreasing over time, and stressed that the implementation led to a need for new skill sets for the drivers and strong work changes for the workers using the scanners (3: 43). However, these changes were learned relatively quickly (5: 100–104).

With respect to the third stage, decision, the complexity and individuality of the innovation decision process becomes especially apparent. It becomes clear that the information-seeking and information-processing nature of the innovation decision process may lead first to adoption and later to rejection or vice versa. The interviews showed that rejection took the form of greater use of the old processes (1: 161), the use of traditional communication channels, especially direct communication even when not needed (3: 53–54), or the observation of the interviewees that other workers were not convinced (5: 82–84). The adoption of the innovation can be linked to the implementation and beginning of use, particularly the overcoming of scepticism (4: 32–33) and an observation of improvement over time (5: 170–172) – they became friends with the innovation.

Within the fourth stage, the workers report changed behavior as one sign of the implementation of the innovation. While supervisors and management decided to implement the innovation, the decision to actually and correctly use the innovation remained with the workers. When the behavior of the workers is considered as one indicator of adoption, it becomes clear that the workers use most of the new control aspects, such as tracing through the app (2: 83–85) and checking co-workers' work (2: 126–131). Also, by using the scanners the workers contribute to the new work process and thereby adopt the new innovation. Here social control mechanism seems to have influenced the decision stage. The interviewees reported that they regularly ask their co-workers to scan items (5: 36–37) and that the tugger-train drivers will not transport produced parts that are not scanned, leaving the worker with unfinished work (5: 106–122).

Lastly, concerning the confirmation stage, two points are clear. First, some interviewees experienced dissonance with the behavior of their co-workers (5: 115–121) or with the hardware of the innovation (3: 215–217), leaving them frustrated with the implementation process. Second, the workers can evaluate the innovation as tried, tested and now approved (4: 32–33) or as an improvement for the whole company and therefore worth the process (3: 202–203).

4.2 Worker's Perception of Empowerment

The innovation decision process has close ties to the concept of empowerment. With respect to Spreitzer's [18] first dimension of meaning, the workers describe a fit between the requirements of their changed work role and behaviors, highlighting the relative advantages of the persuasion stage of Rogers' model. In relation to their own ideals and standards, they changed their behavior to adopt the relative advantages of a new predictability and enhanced work environment (4: 21), improved process flow, and reduced workload (3: 19). In addition, they stated that work goals and purpose changed owing to the beliefs of their co-workers, which triggered a process of learning (4: 9). The role and value of the social system of co-workers and supervisors also gained relevance owing to the exchange of social support and improvements for the innovation (3: 41). All in all, concerning the dimension of meaning, the workers emphasized the relative advantages of the innovation, stressing that the innovation fits the requirements of their work role better than the prior systems (3: 13, 4: 19). The interviewees described this fundamental change in collaboration concerning the meaning or value of their work goal in various ways: "I tried to explain that they had to see the greater good" (2: 54), "it must be turned just like a carousel" (5: 112–123), and "it needs to be a real rhythm" (3: 197–198).

With respect to competence, the workers noted that next to the relative advantages they experienced (4: 19), the reduced complexity and new transparency enriched their belief in their capability to perform activities with skill (3: 13; 5: 12). Additionally, the perceived degree of novelty and the role of the social system arise simultaneously and in connection with the workers' belief that they are performing with skill. The perception of the innovation as a further development that connects current challenges to past experiences (4: 9) leads to a second form of learning process (5: 36) and the acquisition of knowledge (2: 12), supporting the workers' perception of competence. The social system supported this development by the feeling of appreciation by co-workers (5: 12) and a new quality of face-to-face communication, now concentrating on improvements rather than accusations (1: 53; 5: 12).

With respect to workers' self-determination regarding the implementation of the innovation, the advantages of the innovation become apparent. The workers stated quite straightforwardly how the innovation gives them the power to oversee their own work and work processes (4: 33) and how they are now able to plan their work day and initiate and regulate actions concerning their work processes. They reported increased ranges of autonomy, just as Spreitzer [18] states, owing to influencing methods, pace, and efforts of their work using the innovation. The reported ranges of autonomy connected to self-determination also become clear as the workers describe that the innovation and especially the software and data analysis regulate their work day more closely than previously (2: 10, 4: 33). The workers also report that they enjoy the reduced uncertainties and have learned beneficial new ways of initiating and regulating their work using the innovation (2:14).

In summary, the workers report that with respect to the innovation, they have an impact on their supervisors, and hence an impact on strategic and administrative outcomes, through reporting suggestions for improvement and an enriched exchange with them (3: 41). Other workers stated in this context that they perceive the developers, their co-workers, as the point of contact (5: 16), supporting their statements of the innovation's impact. The workers also refer to changes in the corporate culture, which supports influencing behavior, and note that they had an influence on the learning process through advising co-workers and had enhanced direct communication when problems arose (5: 16). Using the app allows them to regulate their own working environment by tracing parts that they made, and they understand their participation in work organization (2: 83–85). In addition, they get an overall view of superior work processes, which allow some of the workers to control the scanner use of others and act on misuse (2: 126–129). The workers thus get a sense of the degree of their influence on the outcome of the implementation of the innovation.

4.3 *Participatory Design of the Innovation Process*

In addition to revealing the connection between the innovation decision process and empowerment, this case study exemplifies the effect of a participatory design on the implementation process. The workers stressed that several aspects of their team cooperation changed. First, improvements were made and implemented together (3: 54–57). In addition, co-workers worked together, connecting directly to their self-determination (5: 52) in that they had higher levels of autonomy to initiate and regulate how the innovation would look at the end. This participatory design can also be linked to workers' competence, especially that of the developers, who reported that they were able to learn new skills and combine these with their experiences into the innovation and its design (2: 2). However, regarding further improvements of the innovation, the workers are somewhat disheartened. While the innovation persists (2: 162–175), at the time of the interviews, further improvements had been postponed indefinitely, even though new improvements are still being suggested. The developers helped to empower their co-workers by supporting them when they suggested the innovation (3: 9), by leading the learning process of their co-workers (4: 9), by being their contact for any issues with the innovation (2: 4), and by actually taking their co-workers' suggestions and integrating them into the process of implementation (5: 44). The developers reported that as agents of the innovation, they had to argue in favor of the innovation and explain the advantages of the innovation to their co-workers (1: 59). These aspects of the co-workers' empowerment also underline the impact and self-determination of the developers. Regarding the participation and inclusion of workers, the interviewees report that various colleagues worked on the innovation and its advancement to achieve improvements together (2: 190–192). They state that they feel like they participate in the development and that especially the developers listen to their concerns and ideas. The workers mention that they communicate ideas and problems more often because they view the developers as responsible for the development of the innovation, even if they know that the management supports the innovation. Within this context, the workers describe quite vividly that the innovation lives because of the continuous development and their engagement.

The results show that changed work organization, accompanied with autonomy and a changed perspective, empower workers by giving them more control and transparency regarding their work and work processes, resulting in greater meaning and self-determination. The interviewees describe their work environment not as restricting them or giving them fewer options, but as giving them greater control over their own work and action. Through scanning, they now document their work process and make it transparent for other workers, demonstrating their responsibility. They therefore increased their self-determination and self-efficacy. Workers can now finish work processes by scanning parts that they have produced and assigning the next step for the part to the next worker who is supposed to work on it. Before the innovation, they were not able to control or trace their own work within the greater production process and evaluate their impact on any outcomes. In this context, the social system clearly influences the use of the innovation itself, as peers regulate themselves and others through being a role model and by speaking up to workers who reject the innovation in a respectful way.

As the case study makes clear, the participatory design not only influenced the development of the innovation but also embedded the innovation the social system of the workers. The participation and inclusion of the blue-collar workers shows the positive influence of the perception of the innovation as well as the influence of working relationships and the collaboration of the workers. The workers describe empowerment as the possibility to contribute ideas with the help of a participatory design and with a higher acceptance of the innovation.

5 Discussion

New technologies still require the cooperation with human workers. In this sense, the aim of this chapter was to analyze how the acceptance of workers can be promoted in the implementation of innovations in organizations. The study provides insights into a case where a digital innovation was introduced into a production workflow and help to understand the process of acceptance, but also highlight changes and impacts on empowerment. The analysis shows the significant perspective offered by the Diffusion of Innovation Theory in the context of digitalization in organizations. Moreover, the analysis shows how digitalization processes in organizations take place and how workers respond. Finally, the analysis reveals that the critical factors influencing innovation diffusion are participatory design and empowerment of the workers. These factors lower the inhibition threshold and promote identification of the relative advantage, the compatibility, the complexity of transition, the possibilities of testing, and the visibility of benefits as critical factors for the workers' acceptance.

The workers in this study were able to actually design the advantages themselves and got greater access to information through peer communication channels, giving them a better chance to understand and learn from peers. In this sense, we identified critical antecedents for individuals' collaboration with digital devices, helping to understand whether they are to become friends or foes.

The case study shows quite vividly that Rogers' [15] innovation decision process can be used to analyze the acceptance or rejection of an innovation that is implemented in a participatory design. The participatory design of an implementation process changes how communication channels are to be viewed. For example, the developers drastically influence all stages as they manage the innovation and the possibilities of participation. In addition, the face-to-face communication of the other workers becomes more active if communication is a possible start of a change of the innovation. In this case study, the workers reported an impact on their learning processes concerning the knowledge stage. The developers as peers formed the training courses of the other workers and answered questions in a low-threshold way. Although Rogers [15] himself highlights the importance of peers in the process, he does not connect the existence of knowledgeable peers in the information-seeking process. The solutions and improvements by and through workers changed the perceived characteristics of the innovation, as they were able to customize the advantages of the innovation to fit their needs, giving greater significance to the relative advantages. Moreover, the information-seeking process of workers is supported by open communication from peer to peer, embedded in the participatory design. This process promotes empowerment since workers experience consistent values and work goals. Moreover, they have the competencies to contribute to organizational processes and the self-determination to initiate and regulate actions. Finally, they perceive to have an impact on organizational outcomes. The following Fig. 3 depicts the interrelation of the relevant concepts discussed.

In organizations, new technologies are usually initiated top-down by executives. However, acceptance of new technologies in the work process and the perception of empowerment are greater if the organizational culture is designed so that the workers participate in the change process and the challenges of the transformation are shaped in partnership with the organization and its workers [15]. This effort can be supported by a participatory implementation of innovations and the consideration of the social system. The direct participation of workers could be encouraged by continuous improvement processes and employee suggestion systems enabling the consideration of specific knowledge about work processes, opportunities, and

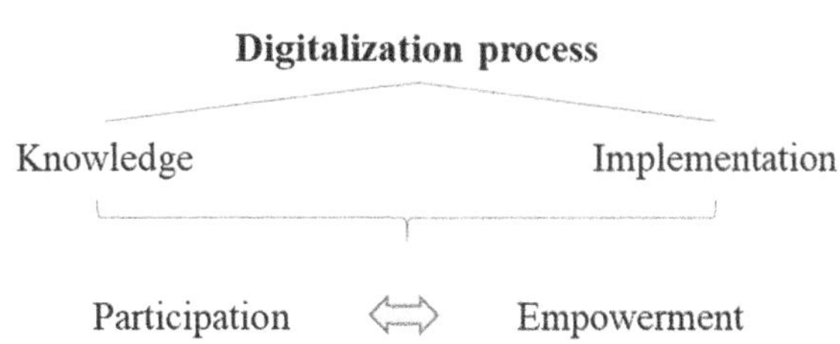

Fig. 3 Participation and empowerment in digitalization processes

shortcomings. In addition, participative leadership helps to support the workers' empowerment [21].

Our findings must be considered in the light of their limitations. First, since we conducted a case study, only one production organization was investigated regarding the implementation of an innovation. The findings therefore relate to a specific organization. Future studies should attempt to replicate the results in other organizations and contexts. Second, while the study has provided valuable insights into the workers' perceptions, the qualitative interviews are limited to a highly specific sample of five production workers within Germany. In a future study, a broader data basis integrating white-collar workers would be insightful. To further elucidate the acceptance of innovations in organizations, the sample could also be expanded in future studies with regard to the perspectives of non-adopters.

References

1. Autor DH (2015) Why are there still so many jobs? the history and future of workplace automation. J Econ Perspect 29(1):3–30
2. Bader V, Kaiser S (2017) Autonomy and control? how heterogeneous sociomaterial assemblages explain paradoxical rationalities in the digital workplace. Manage Revue 28(3):338–358
3. Bonin H, Gregory T, Zierahn U (2015) Übertragung der Studie von Frey/Osborne (2013) auf Deutschland. Forschungsbericht. Bundesministerium für Arbeit und Soziales. Berlin. Retrieved September 09, 2019, from BMAS: ftp://ftp.zew.de/pub/zew-docs/gutachten/Kurzexpertise_BMAS_ZEW2015.pdf
4. Cappelli P (1999) The new deal at work: managing the market-driven workforce. Harvard Business School Press, Boston
5. Frey CB, Osborne MA (2013) The future of employment: how susceptible are jobs to computerisation. OMS Working Paper, Oxford Martin School, Oxford.
6. Gunasekaran A, Ngai EWT (2014) Expert systems and artificial intelligence in the 21st century logistics and supply chain management. Expert Syst Appl 41(1):1–4
7. Hackl B, Wagner M, Attmer L, Baumann D (2017) New Work: Auf dem Weg zur neuen Arbeitswelt. Springer Gabler, Wiesbaden
8. Klumpp M, Hesenius M, Meyer O, Ruiner C, Gruhn V (2019) Production logistics and human-computer interaction—state-of-the-art, challenges and requirements for the future. Int J Adv Manuf Technol. https://doi.org/10.1007/s00170-019-03785-0
9. Laloux F (2014) Reinventing Organizations: A Guide to Creating Organizations. Nelson Parker, Brussels
10. Mayring P (2000) Qualitative content analysis: qualitative social research, 1. Retrieved September 09 from https://www.qualitative-research.net/index.php/fqs/article/view/1089/2385
11. Mazmanian M, Orlikowski WJ, Yates J (2013) The autonomy paradox: the implications of mobile email devices for knowledge professionals. Organ Sci 24(5):1337–1357
12. Montreuil B (2011) Towards a physical internet: meeting the global logistics sustainability grand challenge. Logistics Res 3(2):71–87
13. Orlikowski W (2016) Digital work: a research agenda. In: Czarniawska B (ed) A research agenda for management and organization studies, Edward Elgar Publishing, Northampton, pp 88–96
14. Phan S, Ballot E, Huang GQ, Montreuil B (2017) Physical Internet and interconnected logistics services: research and applications. Int J Prod Res 55(9):2603–2609
15. Rogers EM (2003) Diffusion of innovations. Free Press, New York

16. Sattelberger T, Welpe I, Boes A (2015) Das demokratische Unternehmen. Neue Arbeits- und Führungskulturen im Zeitalter digitaler Wirtschaft. Haufe, Freiburg/München
17. Schraeder M, Swamidass PM, Morrison R (2006) Employee involvement attitudes and reactions to technology changes. J Leadersh Organ Stud 12(3):85–100
18. Spreitzer GM (1995) Psychological empowerment in the workplace: dimensions, measurement and validation. Acad Manag J 38(5):1442–1465
19. Wieland A, Handfield RB, Durach CF (2016) Mapping the landscape of future research themes in supply chain management. J Bus Logistics 37(3):205–212
20. Yin RK (2014) Case study research: design and methods. Sage, Thousand Oaks
21. Yukl GA, Becker WS (2006) Effective empowerment in organizations. Organ Manage J 3(3):210–231

Caroline Ruiner is professor for sociology at the University of Hohenheim in Germany. She studied sociology and management science at Goethe University Frankfurt, received her doctorate in sociology at the University of Augsburg and habilitated at the Faculty of Social Sciences at Ruhr University Bochum. Her research focuses on new work and new employment relationships and the effects on individual and organizational levels as well as with regard to industrial relations.

Christina E. Debbing is a research assistant at the Institute of Education, Work and Society at the University of Hohenheim in Germany. She studied integrated social sciences at Technische Universität Braunschweig and social sciences within the study program management and regulation of work, economy, and organization at Ruhr University Bochum. In relation to the sociology of work, economics, and organization, her research focuses on participation and social support in digitalized work contexts and innovation diffusion.

The Logistics Labor Market in the Context of Digitalization: Trends, Issues and Perspectives

Andrea Bottalico

Abstract This chapter aims to provide a general picture of the current transformations in the logistics labor market in the light of digitalization and automation processes, affecting working conditions, job profiles, training systems, and skills required. The chapter reflects on the implications of labor dynamics along the maritime-logistics chain, and on a reconfiguration of the human resource management in the future, due to the ongoing structural changes and dynamics that suggest the brink of a new technological leap in the capitalist system. The digital transition related to the ongoing crisis with the social and employment consequences deriving from the technological change in the production system, could speed up the changing processes related to digitalization and automation on the logistics labor market. In light of these new scenarios, this chapter seeks to outline an explanatory framework for analyzing the logistics labor market in the context of digitalization, with its peculiarities, challenges and trends.

Keywords Logistics labor market · Digitalization · Automation · Port labor

1 Introduction

Logistics, intended as a management strategy of optimized flows, has now established itself as one of the strategic aspects of corporate businesses, introducing a new paradigm in the business economy. The organizational and transport complexity of a supply chain is directly proportional to the increasing structuring of the firms in "networks" and the fragmentation of global value chains. As Veltz [15] reiterated, the modern economy generates more and more synchronization needs, which adapt to the general trend of reduction of production time cycles. In a historical era marked by humanitarian crises, relative intensifications of migration flows and pandemics, the transport of goods presents itself as one of the fundamental sectors for the global economy. On the other hand, the health emergency resulting from the spread of

A. Bottalico (✉)
University of Milan, Milan, Italy
e-mail: andrea.bottalico@unimi.it

M. Klumpp and C. Ruiner (eds.), *Digital Supply Chains and the Human Factor*, Lecture Notes in Logistics,
https://doi.org/10.1007/978-3-030-58430-6_7

Covid-19 has increased the awareness that, in a context of global value chains and strong interdependencies between economies, a shock that hits one of the links in the chain is enough for a systemic impact. Once the emergency is over, according to some observers, the de-globalization process will accelerate [1]. Firms will consider most likely the lessons learned about the potential breakdown and disruptions of global supply chains due to these shocks, and will reflect, on a reconfiguration of the business models adopted so far. However, if some observers affirm, on the one hand, the fragility of global value chains and the trend of de-globalization, on the other hand, they will foresee a growth in the logistics sector (despite the slowdown in international trade), freight forwarders and couriers (for the growing home delivery request). All these dynamics suggest that we are on the brink of a new technological leap in the capitalist system, and that the digital transition related to the ongoing crisis, with the social and employment consequences deriving from the technological change in the production system, could even speed up these structural processes. The impact of these ongoing dynamics on the logistics labor market in the context of digitalization remains a field to be explored [5, 7].

Furthermore, in recent years, many changes occurred along the logistics transport chain. The expansion of the shipping companies towards the port services market and the inland logistics market (transport, shipping and logistics platforms), for instance, is boosting the competition along the whole of the logistics network. To understand the new challenges of the digitalization related to the logistics labor market, it is necessary to consider its totality. In light of these new and unprecedented scenarios, this chapter seeks to outline an explanatory framework for analyzing the logistics labor market on the basis of the research conducted so far and some recent studies. In the next section, key concepts necessary to understand the perimeter of the analysis will be defined, followed by a set of studies that focus the attention on labor regimes along the maritime-logistics chain, and a final section which, in the wake of this analytical approach, will focus on the variety of labor regimes in the logistics system of freight transport in Italy, with its peculiarities, challenges and trends.

2 The Logistics Labor Market

The professional profiles that we find in the logistics labor market depend in part on the boundaries that are established to delimit it. Dock workers, porters, seafarers, road haulers, airport workers, train drivers, but also couriers, riders or drivers involved in the last mile of urban logistics—all these profiles operate within that complex panorama that we call logistics, with different contractual and working conditions from each other. Some of these profiles pertain to the transport of goods by land, sea or river, air or rail; others concern the movement and handling of goods in warehouses or distribution centers, in port or airport terminals, or in intermodal railway terminals. To these professional profiles we might add the increase of technical staff, the middle managers (white-collars), the IT specialists of the companies and freight forwarders, the logistics operators on behalf of third parties (3PL). As can be imagined, the profiles are heterogeneous as are the skills (high or low), the composition of the workforce, the work environments, the employment conditions and the forms of representation.

Along with that, the scientific literature acknowledges that the future innovations in logistics and supply chain management are driven by technological, regional, economic and social changes. The logistics and transportation industry has become a knowledge-intensive service industry, similar to other sectors such as banking, insurance, and engineering [10]. The digital transformation has reconfigured both businesses and personal lifes in recent decades, with an unprecedented growth in the production and usage of data.

Moreover, the technological innovation introduced in the organization of labor, as well as the automation processes, represent a sensitive issue for social partners. Automation processes, for instance, produced unavoidably a contraction of the number of dockworkers in the port industry. Since the 1960s, European ports have experienced a contraction or stagnation of workforce. Technological innovation, increased containerization, intermodal transport, and the integration of container terminals in global supply chains are just some of the elements that have characterized this process of transformation along the maritime-logistics chain [13]. With the advancement of automation, the organizational structures, the professional and social status of the dock labor have deeply changed: less respected as a very heavy job and less attractive among young people. These dynamics are changing the nature of jobs. The question of professional skills in this segment of the logistics chain is relevant. The main issue is the erosion of jobs caused by the automation processes. Some professions may gradually lose importance, but it seems that job loss is not the most relevant problem. If some professions become obsolete due to the introduction of new technologies, other professions will be created and these will at least partially compensate for the loss of jobs. Along this line, trade unions in Italy speak about the individual right to digital literacy and the adjustment of skills.

To estimate the consistency of logistics occupations in all economic sectors, some studies uses the Ateco Istat sub-sector of "logistics as a service" as a reference. An element to consider at the outset therefore concerns the definition of "logistics". According to the Italian Logistics and Supply Chain Management Association (AILOG), the term defines the set of organizational, managerial and strategic activities that govern the flow of materials and related information in the company from the origins to the suppliers up to the delivery of finished products to customers and post-sale service. This definition suggests that talking about logisticis in general is not correct, because there are as many logistics sectors as productive sectors. The automotive logistics will be different, for instance, from the pharmaceutical logistics, and so on.

Zijm et al. stress that there exist different definitions for supply chains, logistics and their operations. Considering the definitions provided by the American Council for Supply Chain Management Professionals (CSCMP), the authors emphasize that logistics to the transportation and storage of materials, parts and products in a supply chain. Logistics therefore includes inbound and outbound processes to and from warehouses, as well as internal and external materials handling, and transport operations. It also includes the execution of services and the transfer of information between the various stages of a supply chain [9].

Another relevant difference to consider is that between logistics intended as a value production tool (as in the German case) and logistics understood as a cost, as is the case in the Italian production system, characterized by small and medium-sized enterprises operating on international markets, that typically outsource this service to third parties (the so-called "Ex Works", [3]). This model has significant consequences not only on the logistics labor market. From such a productive model, which relegates the logistics activity and its labor to a mere cost to be cut, emerges the recourse, above all in warehouses and logistics platforms, to a contracting and subcontracting mechanism that leverages on the peculiar Italian cooperative system for the recruitment of cheap labor, in addition to the violations of the contractual conditions.

3 Labor Along the Maritime Logistics Chain of Freight Transport

A relevant approach to observing the logistics labor market is that which considers the entire port-logistics cluster to observe the variety of the working regimes that constitute it. The assumption behind this approach concerns the paradigm shift that occurred with the logistics revolution [4]. Notteboom and Rodrigue [12] underline the process of decentralized distribution within regional networks, in which the efficiency derives from the high levels of integration with the internal transport distribution systems. Studies on port spaces have also undergone an epistemological change in the conceptualization of the port, from a single fixed spatial entity to a network of terminals operating on different scales. The scientific literature has focused its attention on the development of maritime networks and hinterlands, thus, glancing at the entire supply chain in the context of globalization and the dissolution of these spaces, which remain territorially embedded on the one hand, and on the other hand, are deterrotorialized through the regionalization process [6].

Furthermore, in recent years, the huge growth of the main shipping companies, gathered in the three alliances that today control the international movement of goods, is gradually expanding towards the port services market, but also of that traditionally carried out by shipping agencies, with further forays into the inland logistics market (transport, shipping and logistics platforms). The main effect of this trend is a competition that takes place along the whole of the logistics chain, generally formed by the maritime activities, the handling of goods in the port space and the transport services in the hinterland [11]. The strategy of the main players in the concentration processes does not seem to be limited to the maritime fraction of the logistics chains, but aims to achieve control of the entire chain in order to exploit the economies of scale generated by the optimization of operations, but also by the reduction of intermediation times and costs.

These three legs (maritime activities, handling of goods in the port space and transport services in the hinterland) are more and more integrated, whereas ports represent the central and essential link of the logistics chain. For this reason mainly, in recent decades ports have experienced a paradigm shift. They have been integrated as links into broader supply chains and global production networks, representing pivotal links within maritime supply chains and global production networks, while also being embedded within specific, path-dependent, spatial and institutional frameworks. The competitiveness of global production networks is mainly determined by the performance of the highly dynamic logistics networks as they link production, distribution and consumption [8]. To understand the new challenges of the logistics labor market in light of the digital transformation, hence, to understand it is necessary to consider the totality of the logistics chain [9, 14].

4 The Structure of the Maritime-Logistics Chain

Mersmaan et al. [11] tried to represent the main features of a maritime-logistics chain in order to understand the new challenges of the overall sector. First, the maritime-logistics chain is a network and the main actors are the cargo owners, the shipping companies, and intermediaries such as agents and forwarders. The terminal operating companies into the ports will depend on the decisions made by those three parties, but the shipping companies are the lead firms in this network and in the port area. It is precisely through the cargo flow along this chain that the value is produced and the competition takes place.

The dotted lines in Fig. 1 indicate alternatives to direct paths that could be undertaken, involving one or more intermediaries. The owner of the goods or shipper chooses a shipping company with or without the mediation of a forwarder, as shown in bold lines. In the reverse case, marked with non-bold lines, the receiver of the goods makes that choice. In turn, the shipping company will opt for a specific route and then for a port of call, either in collaboration with the shipper or independently of them. On arrival at the port, with or without an agent's mediation, the shipping company chooses a terminal operator. The final stretch of the route requires the choice of a logistics operator and of intermodal transport in the hinterland. This decision might be made by the shipper (in bold), the receiver of the goods (not in bold), or the shipping company. Distribution centers (in dotted boxes) can be used in the land stretch. The main issue is to organize this complex context so that market forces can ensure the flow of unhindered goods through the logistics chain in the most efficient way.

In the last years, the shipping companies have focused attention on terminal operators and hinterland transport services. Their strategic alliances and ownerships (partly or entirely) of terminal operating companies is a rather widespread phenomenon. In the European Le Havre—Hamburg Range (called Northern range),

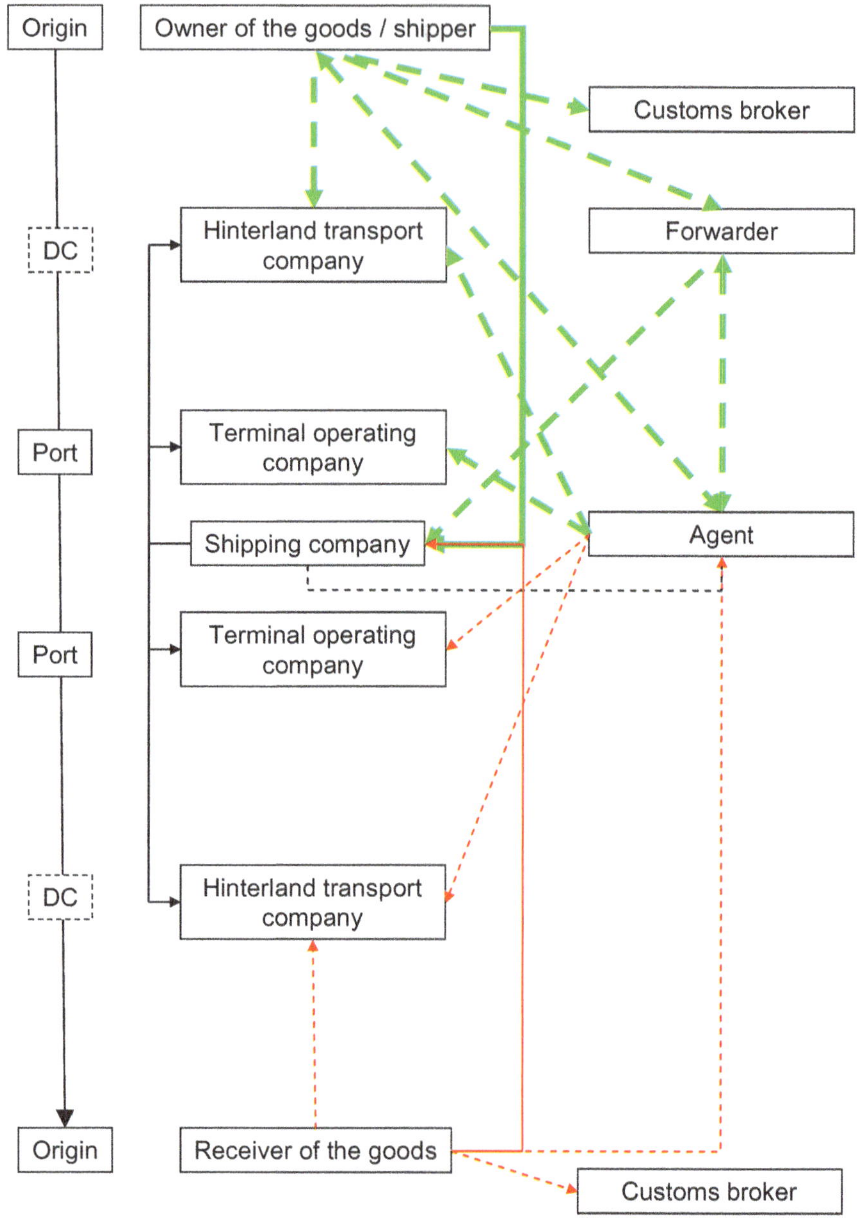

Fig. 1 Structure of the maritime-logistics chain [11]

for example, where the most efficient ports in Europe are located, the main players are global terminal operators and market-leading shipping companies, which cooperate in most cases. Seven main terminal operators are located in twenty terminals of which one (the subsidiary APM Terminals) is directly owned by a shipping company (Maersk). In thirteen of the twenty terminals there is a form of cooperation with a shipping company (vertical integration or joint ventures), while the other terminals are "independent" (Table 1).

Taking into consideration, for example, the 2 M alliance (Maersk and MSC) it is possible to assess the effort of each of the two partners in the conquest of spaces within the port arena. APM Terminals (APMT), until 2017, was present in 69 countries with terminals located in 73 ports and 140 hinterland transport services. Being owned by the shipping company A.P. Moller Maersk, APMT is vertically integrated. In addition, Maersk line also offers forwarding and supply chain management services through its Damco company. In this way, Maersk Line is present in the entire supply chain. Terminal Investment Limited Sàrl (TIL) is the second terminal operating company owned by a shipping company (MSC) and active in Northern Europe. TIL is a terminal operator founded in 2000 to guarantee mooring and capacity for ships of the MSC shipping company. The group manages and operates 36 terminals in 24 countries, most of them in joint ventures with other terminal operators (situation in 2017).

Nevertheless, the compatibility between national regulations and neoliberal policies and regulations at European level has been a strong influence on port services and in particular on the variety of port labor systems and schemes. The aim of European institutions in recent years has been to liberalize port services, including port labor, according to the principles of the European Treaty on freedom of establishment and freedom to provide services (Article 49 of the TFEU, Treaty on the Functioning of the European Union). On the other hand, national port labor systems and schemes in Europe, in addition to other variables, very often tend in the opposite direction. In 2016, for instance, a port labor reform took place in Belgium after the infringement procedure sent by the European Union to the Belgian government, concerning the incompatibility of the port labor system with the principles of the Treaty. The same process took place in Spain in 2017 and in Italy before the port reform in 1994. The infringement procedure in Belgium started from the complaints sent to the European Commission by a multinational cargo handling company, which is involved mainly in logistics activities in the port of Antwerp (one of the most efficient ports worldwide in terms of productivity, throughput, etc.), and by several interim agencies. After months of talks, the social partners proposed a process of port reform to the European Commission to be implemented in the coming years. The solution proposed by the Belgian government was positively assessed by the Commission in May 2017 and the infringement procedure was withdrawn, which concerns the "freedom of establisment". Although the port of Antwerp is mostly carried out by multinational companies locally situated, the infringement procedure sent by the European Commission to the Belgian government concerning the organization of port labor is derived from the incompatibility between the national law and the article 49 TFUE, concerning the "freedom of establishment".

Table 1 Cooperation among terminal operators and shipping lines (Northern range, 2017)

Terminal operator	Name of terminal	Location	Cooperation with shipping line
Eurogate	Container terminal Hamburg	Hamburg	
Eurogate	CTB (container terminal Bremerhaven)	Bremerhaven	
Eurogate	NTB (North sea terminal Bremerhaven)	Bremerhaven	Maersk
Eurogate	MSC gate	Bremerhaven	MSC
Eurogate	CTW	Wilhelmshaven	Maersk
PSA	Noordzeeterminal	Antwerpen	
PSA	Europaterminal	Antwerpen	
PSA	MPET	Antwerpen	MSC
DP World	Antwerp gateway	Antwerpen	Cosco pacific, CMA CGM,ZIM
DP World	RWG (Rotterdam world gateway)	Rotterdam	APL, MOL, HMM, CMA CGM
DP World (GMP)	Terminal de France	Le Havre	CMA CGM
Terminaux de Normandie (TN)	TNMSC	Le Havre	MSC
Terminaux de Normandie (TN)	TPO	Le Havre	
HHLA	CTA (Container terminal Altenwerder)	Hamburg	Hapag-Lloyd
HHLA	CTB (Container terminal Burchardkai)	Hamburg	
HHLA	CTT (Container terminal Tollerort)	Hamburg	
Hutchinson Ports	ECT Delta terminal Rotterdam	Rotterdam	MSC
Hutchinson Ports	ECT Euromax terminal Rotterdam	Rotterdam	Cosco Pacific
APM	APM Terminals Rotterdam	Rotterdam	Maersk
APM	APM Terminals Maasvlakte II	Rotterdam	Maersk

One of the common peculiarities of port labor deals with the uncertain dynamism of the maritime traffic. Dock work depends on the relentless and unpredictable rhythm of the goods. The management and the governance of port labor are particularly important with regard to the application of the basic rules of the European Treaty. The organization of dock labor schemes is in fact regulated and mostly subject to Treaty rules on competition at European level. The application of these principles to port labor systems is one of the key debates in the port sector. The debate between social partners at European level refers to the forms of "protection" of the external pressures to which port labor is subject and the "restrictions" to the free market.

5 Value Creation and Employment in the Logistics Transport Chain

Although various types of goods are handled in a universal port, the container—the management, storage and transport unit—represents the most important "commodity" in statistics due to the added value it creates, the economies of scale it produces, the supply chain and the multiplicity of professional figures that it involves from origin to destination. In this regard, a study conducted by Blomme [2] tries to answer the question of what the economic contribution of the different sub-sectors in the port area of Antwerp is. To answer the research question, according to Blomme it is necessary to calculate the added value created by the ports taking into consideration the intermodal structure of the maritime-logistics chain. From this perspective, it is possible to identify the main activities and the added value that they create along the entire transport chain, consisting of docking operations, goods handling activities in the port and transport operations in the hinterland. All these activities along the maritime-logistics chain generate, to a greater or lesser extent, added value and are closely interconnected with each other.

In this perspective, the different components of docking in port and port handling activities are considered, plus the added value created by shipping agents, forwarders and shippers, and finally the added value associated with the transport of goods in the hinterland. The added value of the entire transport chain relating to ports is far greater than the only added value of the port handling activity at the terminals. Another interesting aspect is the observation of activities with little or high added value and employment. Logistics-related services, for example, create greater employment and high added value. Also the context of industrial relations should be considered in this context. In the Belgian case, for instance, there is a strong culture of social dialogue, an important role of work councils, joint committees and subcommittees. Working conditions are determined by means of collective bargaining agreements concluded within these joint committees. Belgium's trade union membership rate is among the world's highest and it is fruitful in terms of the conclusion of agreements.

6 Port Labor, Railway Work and Intermodal Transport in Italy Between Digital Transitions and Organizational Changes

In the wake of this approach, if we look at the absolute numbers of the Italian freight transport logistics system (Table 2), port labor and railway work are residual within the logistics chain, both for employees and for wealth generated. Beyond the economic value generated, port labor and railway work, however, have a strategic value within the logistics chain because they cover two obligatory passages, the first linked to the transfer of goods (port) from land to sea, the second tied to the economic and environmental sustainability of the traffic increases caused by the growth of the tonnage of ships (intermodality sea-rail).

All this requires advanced skills, professional training and specializations, both in port and in the exchange and forwarding on the rail carrier, in particular in light of the digital transition and the quick upgrading of the skills required. However, some recent phenomena are questioning these profiles. On the one hand, driven processes of robotization and digitalization of the port cycles reduce the human contribution and increase the need of qualified personnel, due to improved technology interaction. On the other hand, the progressive horizontal integration between the main shipping companies worldwide and vertical integration between the latter and the most important terminal operating companies have contributed to redesigning the port business by redefining roles and competences of each actor involved. The increase in the size of the ships has resulted, for example, in addition to a reduction in the number of ports of call, in a greater concentration of the employment. The consolidation of a small number of alliances of the main global shipping companies, moreover, leads to an increase in traffic peaks and intermittent labor demand within ports. In light of the robotization processes, the technological innovations and market integration, rather than excluding labor due to an increase in automation, the most acute risk that must be faced in the future concerns the potential trivialization of human action and its marginalization. Along this line, the digital transition in the logistics labor market shows the key role of professional training due to the new job

Table 2 Ranking of logistics jobs (2019)

	Value added (Millions of €)	Employees
Port logistics, auxiliary activities and maritime transports	2.213	27.611
Rail freight transport	69	7.500
Road freight transport	10.571	309.153
Air freight	40	696
Cargo handling	3.119	108.162
Other transport-related activities	4.015	88.043
Postal services and courier activities	7.625	161.240

profiles and competencies required. Whereas in the past many blue collar jobs in logistics required a basic school education and rudimentary qualification levels, it is clear that these jobs now require increased competences. The same is true for white collar jobs in logistics [10]. The market for freight transport, logistics and shipping services is impressive from the point of view of the activities developed, but not very effective from that of the economic results obtained. Organizational models and poorly advanced technological systems limit the performance of companies in the sector characterized by irregular and poorly transparent forms of labor that fuel pulverization and depress the sector's ability to create value from service activities.

The port segment of the maritime-logistics sector has seen worldwide shrinking of the workforce. In Italy, between 1983 and 2001, over 20,000 workers of who were then called "Compagnie Portuali" left work through early retirement measures. The workforce within the Italian ports has changed to the point that today the average age of workers in ports has significantly decreased. The interesting fact is that from 1983 onwards the labor force in ports first contracted, to then reacquire the consistency of the past years. The main change occurred therefore in the organization of work, but not only. Other changes relied on the professional profiles, the new skills and the training required, the increasing flexibility due to the labor peaks, etc. In few words: less muscles, more brain. Most likely, the future of port labor will see more technicians specialized in new automated systems.

The reconversion of the workers employed in the Italian ports did not serve as much to reduce the staff as to articulate the distribution of the latter differently between employees directly employed by the terminal operating companies and those employed to port activities with the intermediation of the port labor pools in place with the Italian port system. This is an important change that is measured by the new rhythms of the transport service and by the reduction of the time available for loading and unloading goods. The costs of managing the ship, both in navigation and in port, have significantly increased, as has the economic value of the damage suffered by goods. The technological innovation has changed the quality and intensity of work in ports, but has not eliminated, let alone reduced to the bone, the need for specialized workers to handle goods from the ship to the shore and vice versa.

In general terms, fields of automation include the maritime segment (unmanned ships), nautical services, cargo information systems, as well as operational automation at port level and supply chain level, with impacts on work and organization. Automation refers to the application of "automatic control", the use of programmable logic controllers (PLC) in machinery, like the automated stacking cranes. It is acknowledged that automation reduces human intervention but not necessarily employment, and that it is a very slow, gradual process. Digitalization is also an important factor in this field. In general, the purpose of automation is to achieve higher throughput or productivity, fewer direct human labor costs and expenses, higher quality or increased predictability of quality, improved robustness of processes or products, a higher degree of accuracy, replacement of operators in tasks that involve hard physical or monotonous work, replacement of workers in tasks done in dangerous environments, etc. The risks of automation concern the reduction of operational flexibility, the reduction of financial flexibility (fixed costs), high initial costs, security and vulnerability. In addition, some tasks cannot be automated, or only at high costs. The risk of operational control produces further the automation paradox: the more efficient the automated system, the more crucial the human contribution of the operators. Humans are less involved, but their involvement becomes more critical.

According to the trade unions at European level, the potential impact on dockworkers depends on the terminal concept (greenfield/expansion/brownfield), increase of volumes and terminal capacity in relevant ranges, current job structure and collective labor agreements, job content and working conditions. However, the job and qualification structures are affected. Automation has an impact on dock labors in terms of skills and job losses, producing a shift from direct to indirect jobs. Health risks may change (also improve), as well as the flexibility, which may increase (more peaks, tight schedules, etc.). The impact on total employment in the next decades is uncertain, while the impact on unskilled/lower skilled workers is expected to be high.

In the case of the port of Antwerp, ICT introduction and automation will boost a polarized labor market, where a lot of middle-paid paperwork jobs will disappear, jobs on the floor will be more and more supported by robotics and data applications, and management jobs will become more and more complex. These major changes are strictly related to training programmes that must include new skills like ICT, soft skills like teamwork and communication, and inclusion of females and non-natives in the port labor market. At terminals, in particular for containers, automation goes slower, but especially ICT introduction leads to the need for more ICT competences. In the non-maritime cluster, digitalization and data sharing will become key issues [7].

7 Conclusions

This chapter aimed at exploring the main trends, challenges, issues and perspectives of the logistics labor market in light of digitalization and the new scenarios, in particular by looking at the port segment of the logistics chain. As we have seen, the technological innovation introduced in the organization of labor, as well as the automation processes, have an impact on the logistics labor market in terms of working conditions, new job profiles and skills, qualifications, number of employees, etc. Port labor and railway work have a strategic value within the logistics chain, but they require advanced skills, in particular in light of the digital transition and the automation processes.

Moreover, digitalization in the logistics labor market shows the relevant role of training systems in the present in order to upgrade the human resource pool involved in the logistics sector [16].

The potential impact on the overall logistics labor market should be considered also with respect to the ongoing crisis produced by the health emergency resulting from the spread of Covid-19, which will speed up a set of digital and automated solutions affecting the job and qualification structures, total employment, professional profiles and skills required, etc.

The analysis of the logistics labor market must take into account the conditions that the intermodal transport has produced over the years. From the industrial districts, which have been the subject of numerous studies and research in the past, greater attention to the logistics districts, transport infrastructures and both digital and material platforms is suitable. These are the workplaces where loading units are transported, opened, emptied ("stuffed and stripped"), refilled and forwarded; where added value and employment are produced. In these places numerous conflicts and tensions have exploded in recent years from which the strategic position of the logistics workers emerges. In this perspective, logistics represents a capillary network, like the arteries of a human body which, as stated by a union leader during an interview, "is the blood that flows in the veins of global capitalism". If we want to have an overall picture of the impact of digitalization on the logistics labor market in light of the ongoing socio-economic changes, an analytical approach capable of observing the logistics chain of freight transport as a whole will be appropriate in order to provide a clear image of the heterogeneous, fragmented and changing occupational panorama that characterizes it.

References

1. Baldwin R, Weder De Mauro B (2020) Economics in the time of Covid-19. CEPR Press, London
2. Blomme J (2014) Value creation in the port of Antwerp: an investigation of sector and cargo-related creation of value and employment. In: Vanelslander T, Sys C (eds) Port business. UPA, University Press Antwerp, Market challenges and management actions, pp 99–118
3. Bologna S Logistica: innovare le imprese, qualificare il lavoro, per vincere la sfida di Industria 4.0. Roma, 22 giugno 2017, Centro Congressi Cavour
4. Bonacich E, Wilson JB (2008) Getting the goods. Port, labour, and logistics revolution. Cornell University press, Ithaca, NY
5. Bottalico A (2019) Towards a common trajectory of port labour systems in Europe? Case Port Antwerp Case Stud Transp Pol. https://doi.org/10.1016/j.cstp.2019.12.003
6. Ducruet C, Lee SW (2006) Frontline soldiers of globalisation: port-city evolution and regional competition. GeoJournal 67(2):107–122
7. Esser A, Sys C, Vanelslander T, Verhetsel A (2019) The labour market for the port of the future. a case study for the port of Antwerp. Case Stud Transp Pol. https://doi.org/10.1016/j.cstp.2019.10.007
8. Hesse M, Rodrigue J-P (2004) The transport geography of logistics and freight distribution. J Transp Geogr 12(3):171–184
9. Klumpp M, Zijm H (2019) Logistics innovation and social sustainability: how to prevent an artificial divide in human-computer interaction. J Bus Logist. https://doi.org/10.1111/jbl.12198
10. Klumpp M, Bioly S, Neukirchen T (2019) Human resource and knowledge management. In: Zijm H, Klumpp M, Regattieri A (eds) Operations, logistics and supply chains management. Springer, S. Heragu, pp 205–229
11. Meersman H, Van de Voorde E, Vanelslander T (2009) Introduction. In: Meersman H, Van de Voorde E, Vanelslander T (eds) Future challenges in the port and shipping sector. The Grammenos Library, Informa, London, pp xix–xxi

12. Notteboom T, Rodrigue JP (2005) Port regionalization: towards a new phase in port development. Maritime Pol Manage 32(3):297–313
13. Sys C, Vanelslander T (2019) Innovations in ports. Technology, information and processes. University of Antwerp, Antwerp
14. Van De Voorde E, Vanelslander T (2014) Trends in the maritime logistics chain: vertical port co-operation: strategies and relationships. In: Vanelslander T, Sys C (eds) Port business, market challenges and management actions. University Press Antwerp, Antwerp, pp 121–240
15. Veltz P (2017) La Société hyper-industrielle. Editions du Seuil et la République des Idées. Vidal, M, Le nouveau capitalisme productif, Parigi
16. Zijm H, Klumpp M, Heragu S, Regattieri A (2019) Operations, logistics and supply chains management: definitions and objectives. In: Zijm H, Klumpp M, Regattieri A (eds) Operations, logistics and supply chains management. Springer, S. Heragu, pp 27–42

Andrea Bottalico is a research fellow at the University of Milan and adjunct professor in organizational sociology at the University of Bergamo. He obtained his Ph.D. in economic sociology and labor studies from the University of Milan. He was a visiting student at the University of Antwerp, Department of Transport and Regional Economics, and C-MAT (Centre of Maritime and Air Transport Management). His research focuses on port logistics, labor relations and employment conditions.

Multicultural Teamwork—Unification of Differences in Digitalized Work Contexts?

Mona-Maria Bardmann

Abstract The following article addresses the phenomenon of increasingly multicultural, organizational teams in digitalized work contexts. It deals with the cooperation amongst and in such teams as well as the resulting formation of relationship networks—called figurations. The research focuses on blue-collar workers. A situational individual case study from the field of storage and distribution work analyzes the relationship networks between team leaders and team members and the extent to which culture influences cooperation. A figurative-analytical perspective—mainly based on the study "The Established and the Outsiders" by Elias and Scotson (1994)— is applied to shed light on the above-mentioned phenomena and to reveal interrelations. Within the framework of a qualitative research design, guided by the grounded theory methodology, eleven interviews and six days of participatory observation were conducted in order to derive a theory of a permeable *unified storage culture* based on elaborated core categories, which leads to a unification of differences in digitalized work contexts. Furthermore, the findings show that the increasing digitalized structures in the observed company supports this. Simultaneously, this *storage culture* impacts the exposed *established outsider figuration*, but is influenced by it in an analogous manner.

1 Introduction

By nature, men are nearly alike; by practice, they get to be wide apart (Confucius [5]).

In times of increasing border-crossing, organizational activities in the course of the advancing digitalization, globalization and the associated increase in multicultural, organizational teams, the topic of interculturality—although not a new one—represents a current and relevant issue. Marquardt and Horvath [23] describe the building

M.-M. Bardmann (✉)
University of Hohenheim, Stuttgart, Germany
e-mail: mona.bardmann@uni-hohenheim.de

M. Klumpp and C. Ruiner (eds.), *Digital Supply Chains and the Human Factor*, Lecture Notes in Logistics,
https://doi.org/10.1007/978-3-030-58430-6_8

of high-performance multicultural teams as one of the greatest challenges for organizations in the twenty-first century and as the only way to succeed in a globalized and digitalized environment. Consequently, the consideration of different cultures appears to be a criterion for success, especially for globally operating organizations. Dealing with different cultures within the company is also decisive for the value-added process, especially with regard to cooperation in multicultural teams. The preoccupation with humanly staffed, multicultural teams also raises the question of the distinctions or determinants through which various relationship networks are formed. Several studies are concerned with the intercultural cooperation of organizational teams. The focus of these studies is mainly on white-collar workers. However, the field of blue-collar[1] workers is disregarded. In this respect, this study focuses on the cooperation of low-skilled[2] blue-collar workers in the storage and logistics sector. Thus, the phrase "all people are equal, but some are more equal than others"[3] borrowed from Orwell's [28] fable "Animal Farm" can be read as a reference to the formation of unequal relationship networks, despite all declarations of equality, and thus serves as a basis for the development of the research question: Which figurations are formed in organizational teams with multicultural staff in digitalized work contexts and to what extent does culture influence the cooperation?

The study "The Established and the Outsiders" by Elias and Scotson [9] as well as essential findings from research on culture and intercultural cooperation [25, 40] serve as a theoretical basis. The empirically verifiable answers to the questions contained in this study base on figurative-analytical research. Accordingly, the present study builds on a total of six days of participatory observation as well as eleven themed, partially structured individual interviews with team members and experts. The grounded theory methodology guides the collection and analysis of the data material in order to derive a theory of a permeable *unified storage culture* based on elaborated core categories. Furthermore, findings of the present study show that the digital structures in the analyzed company favor this development. Simultaneously, this *storage culture* impacts the exposed *established outsider figuration*, but is influenced by it in an analogous manner.

The following chapter divides into eight sections. Following the introductory remarks of Sect. 1, which outlines the structure and content of the study, Sect. 2 presents an examination of figuration sociology and the relevant conceptual delimitations in the style of Elias. Furthermore, the presentation of the study "The Established and the Outsiders" by Elias and Scotson [9] follows in order to apply it to the organizational context. Section 3 provides a brief overview of the concept of culture. In addition, Sect. 4 discusses the state of the art of relevant studies in order to identify research desiderata, to clarify the relevance of the research questions and to situate the study in the organizational discourse. Section 5 describes fieldwork and

[1]White-collar workers are the employees/managers, whereby blue-collar workers are workers [27, p. 21].

[2]Low-skilled work includes activities that do not require special professional qualifications and can be carried out after a short period of education or settling in [14, p. 3].

[3]Original wording: "All animals are equal, but some are more equal than others" [28].

data collection. The empirical material is composed of eleven guided interviews as well as participatory observations in the investigated company. The evaluation and interpretation of the interview transcripts carries out with the help of the methodology mentioned above. The resulting core categories and the theory of a *unified storage culture* derived as well as the figuration inherent in this culture are discussed in more detail in Sect. 6. This characterizes the core of the present study. Section 7 is devoted to the discussion of the study results, while Sect. 8 compiles the results and provides an outlook on the possibilities for further connections and associated research questions.

2 Basic Study: Elias and Scotson—The Established and the Outsiders

The task of figure-analytical research is to represent subjective experience by the actors and behaviors within a network of relationships in such a way that the essential characteristics of the underlying figuration structure are understandable [43]. According to Baur and Ernst [1], figurative sociological analysis in this context focuses on social (macro) and individual (micro) processes, but also on network of relationships (figuration) as a whole.

Elias and Scotson [9] substantiate the theoretical assumptions on figuration and power or balances of power in their figural-analytical study "The Established and the Outsiders". Elias does not understand the human being as an individual separated from society and the latter not as a human-free system. The basic idea is that figurations represent networks of human relationships that are interdependent with one another. Through this interdependence, which can also be described as entanglement, the concept of figuration is distinguished from the concept of interaction [41, p. 138]. It is not the mere accumulation of individuals that is referred to as figuration. The indissoluble intertwining of the actors in a dynamic process characterizes the term (ibid.). The figurations thus do not represent a static entity, but a constantly changing, dynamic process. According to Elias and Scotson [9], a "compelling force" emanates from the figurations:

> There is a certain abhorrence against the idea that 'societies' or to put it less equivocally the configurations which individuals form with each other exercise some power over the individuals which form them and limit their freedom. Yet whatever our wishes may be, looking simply at the available evidence, one cannot get away from the recognition that configurations limit the scope of the individual's decisions and in many ways have a compelling force even though this power does not reside as it is often made to appear outside individuals, but merely results from the interdependence between individuals [9, p. 172].

The concept of figuration takes into account not only relational aspects, but also the continuous possibility of change—the dynamics—of figurations. Because of their fundamental interdependence with each other, people always group themselves together in the form of specific figurations. These patterns remain even when a society disintegrates. The coexistence of people in societies always has a very specific form, even in chaos, in disintegration, and in the greatest social disorder. This is what the term figuration expresses [10]. Elias considers the changeability of human beings as natural: It belongs to the peculiarities of human beings that they are by nature changeable in a special way [10, p. 139]. He points out that he used the concept of figuration, which people form together, to describe an image of human beings. He understands this as a conceptual tool appropriate to reality, which avoided the compulsion, that is much discussed in sociology, to regard the individual and society as two different, antagonistically opposed figures [10].

Central to Elias' figurative sociology is also the concept of the power that individuals have over one another through their mutual dependence. Through the described dynamic process character of the human networks of relationships, the power inherent in them is not static per se, but also dynamic [13]. For Elias, power is therefore not an amulet that one person possesses and the other does not possess, it is a structural characteristic of human relationships—of all human relationships [10]. The core of interpersonal relationships represents these balances of power, which are indicators of the reciprocal relationships of dependence in which individuals find themselves. Consequently, there are no one-sided dependency relationships. The inequality of opportunity reflects in the everyday power and competition between people: It is quite conceivable that someone who appears to be relatively powerless one day may experience an increase in power the next day and thus be more powerful than before. The balance of power is therefore not static, but changeable. Power is consequently a social process for Elias. The basis for these assumptions is, among other things, the study "The Established and the Outsiders" by Elias and Scotson [9], which is presented below.

The study describes the formation of groups in different contexts that can be defined as 'superior' (established) and 'inferior' (outsiders). It is an empirical-theoretical study conducted by the researchers in 1958 and 1959. The authors refer to it as "configurational theory in action" [9, p. 171]. Figuration is thus not exclusively a conceptual figure of thought, but also an empirical paradigm [31]. The researchers use principles of the grounded theory methodology in their survey and evaluation [9].

The studied population of a small English suburban community, which the authors call Winston Parva, differs marginally due to its socio-structural characteristics. The figuration arises along the distinction between old-established and newly arrived families. Elias and Scotson [9] describe this figuration as *established outsider figuration*, with the old-established being among *the established* and the newcomers among *the outsiders* [41, p. 138f., 9]. In this context, a status ideology crystallizes, in which *the established* classify themselves as better, more civilized and more likeable people. *The established* subsequently stigmatize *the outsiders* as not belonging and differently (ibid.). The authors highlight various determinants that help to create and maintain power differentials and thus contribute to maintaining the social separation of the two groups [9, 31]. Analogously, they may be regarded as mutually reinforcing aspects.

The constituting factor for differentiating the groups is in turn the length of residence at the location [9]. The long-established residents share a common past and a consistent canon of values and norms. They have known each other for two or three generations. The cohesion in the group is—in contrast to that of the immigrants, who are strangers both to the old-established and to each other—correspondingly great, they have a higher cohesion potential (ibid.). In order to maintain group identity, they try to protect themselves from the immigrants who they perceive as a threat [9]. The different lengths of stay accompanied by a geographical and spatial seperation of the groups prevent contacts that exceed the professional necessity: "As outsiders are felt to be anomic, close contact with them threatens a member of an established group with 'anomic infection'" (ibid., p. 24). This infection needs to be avoided at all costs. The group punishes deviations from the norm as taboo breaks. The own group sanctions attempts to approach members of the other two groups by subjecting the members to strong social control. In addition, it is necessary to verbally exclude them from the group by "blame-gossip" (ibid., p. 40). Moreover, Elias and Scotson state that the status ideology is one of the most important aspects of the *established outsider figuration* and express in the construction of the foreign and the self on the side of the established. One's own group continually classifies as the better one, while the members of the other group "lumped them all together as people less well bred." (ibid., p. 16). Thus, no more distinctions are made between individuals. The group as a whole is stigmatized as unreliable and dirty (ibid., p. 125 ff) in order to establish an anomic minority which makes it possible to legitimize the social superiority of one's own group [31]. In this situation the less strong members of the group absorb the disgrace as their self-image, which leads to their further degradation [9]. They point out that if you assign a bad name to a group, sooner or later the group will conform to that name. At this point, the researchers accentuate that the actors are trapped in a double-bind (ibid., p. 31 f). *The outsiders* are indeed dependent on the toleration by the group of the established, but the established group needs the presence and the assigned characteristics of the outsider group for its self-construction. The established group thus needs *the outsiders*. The selective distribution of positions of social power is another factor of subordination. The established group occupies important offices in the community through the exclusion of the outsider group. Thus, power differentials arise from the potential for cohesion of the established group.

According to Elias and Scotson [9], the described figuration of the established outsider relations shows recurring regularities and divergences. Nevertheless, the figurations, for all their changeability, have some common features: In essence, one group always creates the figuration by excluding another group from its status and power, creating a monopoly for itself. The exclusion can be strong or weak. One aspect to be emphasized, for example, is the passing on of knowledge (as a means of power): Communication, knowledge transfer—from group to group or from individual to individual—and accordingly the refusal to transfer knowledge never only affect the cognitive side of human relations, but also always the power relations. In addition, the subgroups are potentially dangerous, which creates a certain fear of each other. The collective self-enhancement then contributes to the integration of the group and thus secures its chances for survival (ibid., p. 45).

Transfer to the Organizational Context

The study described above considers organizational contexts only superficially. Although members of both groups work partly in the same factories, under the same conditions the researchers deny these contexts a socially integrating effect that goes beyond the organizational needs and do not investigate them further [31]. Against the background, that the organization in the present contribution is to be viewed from the perspective of the *established outsider figuration*, it seems necessary to shed light on the characteristic framework conditions and the differences to the original study by Elias and Scotson [9]. In the specific case of the present study on intercultural cooperation in organizational teams in digitalized contexts, the focus is on the analysis of relations in and between three different teams (logistics and storage sector) of a company with an international workforce. With the help of the empirical paradigm of figuration described above, it is possible to shed light on different, individual and group-related decision-making and action options [13, p. 23]. A common feature of the framework conditions of a community and an organization is, for example, the hierarchical structure. The offices of pastor and mayor are, for example, positions of high status and social power. According to Elias and Scotson, the position of power is secured by the occupation of office within the community with representatives of the established group. In an organization these positions of social power represent the representation of employees' interests (works council) and—in the present case—by the team leaders and the storage management [13]. In the local company, the team leaders take over the management and coordination of several selected teams. It is worth mentioning that in this company there is no more stock management. The team leaders therefore no longer have a direct superior. The next management level is the plant manager. Another special feature of organizations is the formation of complex patterns of diverse, heterogeneous sub-groups through the interaction of various aspects of discrimination, such as age, gender, origin and cultural ideas, ownership, qualifications and work specialization based on the division of labor [31]. Since management can use formal authority to influence all other groups of participants without being able to influence them in the same way, it forms a special group of actors. Through the possibilities of influence, it is possible to shape the figurations between the participants (ibid).

3 The Concept of Culture

Research on and with multicultural teams and their intercultural cooperation requires a prior examination of the concept of culture, as it opens up a multitude of possible definitions. In everyday language the term is extremely familiar and is often used inflationary in various central social issues. Especially in the social sciences the use of the term is very heterogeneous [38, 39, 44]. Different definitions will be briefly considered in the following in order to highlight a suitable specification for this contribution. In the strictest sense of the word, the concept of culture stands in contrast to what nature prescribes. For example, nature has dictated that man must drink in order to survive. However, what, how and where he drinks can be freely shaped [34]. Consequently, culture is always a product of various forms of design and refers to what has been learned and created and thus to ways of acting and thinking, but also to materially created things [34]. A comprehensive definition, which serves as the theoretical starting point for the present study, is that of Kroeber and Kluckhohn [20]. The authors have analyzed more than 150 different views of culture and subsequently worked out a summarizing definition:

> Culture consists of patterns, explicit and implicit, of and for behavior acquired and transmitted by symbols, constituting the distinctive achievement of human groups, including their embodiments in artifacts; the essential core of culture consists of traditional (i.e., historically derived and selected) ideas and especially their attached values; culture systems may, on the one hand, be considered as products of action, on the other as conditioning elements of further action (ibid., p. 181).

A more recent definition by the cultural anthropologist Maletzke [22] is similarly comprehensive. He considers culture essentially as a system of concepts, beliefs, attitudes, an value orientations. These can be found in the behavior and actions of individuals as well as in their spiritual and material products. In short, according to this concept, culture represents the way humans live and what they do with themselves and their world [22]. Accordingly, both views take up cognitive, affective and behavioral aspects and their underlying values as facets on the one hand, while on the other hand, they list the historical perspective and the influence of symbols as a possibility of transferring cultural achievements as dynamic facets [30]. Thomas [40] suggests a practical definition, albeit not covering all aspects, against the background of dealing with improving cooperation between people from different cultural backgrounds. According to Thomas [40], culture is a universal phenomenon. All people live in a specific culture and refine it. Culture structures a field of action specific to the population, ranging from objects created and used to institutions, ideas and values. Culture always manifests itself in an orientation system typical of a nation, society, organization or group. Specific symbols form this orientation system and it is passed on to the next generation in the respective society, organization or group. The orientation system defines defines their membership in the society or group for all members and and enables them to cope with their own environment (ibid.). Following Moosmüller [25], the definition of culture comprises two further essential aspects in addition to that of the orientation system, which provides standards for action: Firstly, culture is understood as a self-constructed network of meaning. Secondly, culture is seen as an individual culture, which influences the perception, thinking and action of individuals as a "mental program" [15] resp. mentality or as "habitus" [4, 7]. Culture is thus not an ever-present entity, but is regarded as socially constructed, complex and dynamic. Beyond that, culture is an internalized construct that is very closely interwoven with personality [25]. The dynamic character of culture immediately shows similarities to the dynamic characteristics of human figurations described above. The action-oriented view seems to be suitable for the present contribution since it primarily intends to deal with the cooperation of people of different origins in teams.

An assumption for the further course of this work will thus be that cultural influences have a decisive influence on the perception, thinking, values and actions of individuals in the respective society.

Regarding the definition of the concept of culture described above, strong similarities to the concept of corporate culture become apparent. Based on the definition of Sackmann [32],[4] it is therefore possible to transfer the results of dealing with culture in general to the corporate culture and thus to the context of the organization. On this basis, the research question can be answered.

The orientation system accentuated by Thomas [40] offers on the one hand possibilities or incentives for action, but on the other hand it defines the conditions for action and has limitations. It thus makes it possible to generate a sufficiently large stock of knowledge about the material and social environment and to gain experience in dealing with it in order to use it appropriately and effectively (ibid.). With regard to the research guiding question, transferred to the organizational context, it is necessary to examine whether this orientation system also functions in the world of employment. The persons, objects, events, procedures, processes and behavioral sequences surrounding the individual can be attributed meaning and significance. The process of creating meaning usually takes place unconsciously and automatically, i.e. without great psychological effort, and is therefore an individual achievement sui generis. Its emergence, however, is not random and arbitrary, but follows a collective, socially binding system of norms and rules that has a culturally mediating effect. In addition to the system of orientation, instruments (theories, methods, rules, norms, etc.) for adaptation and innovation are acquired in the course of the socialization process, with the help of which the individual can orientate and find his way [40]. These characteristics are also transferable in the broadest sense to the structures in the researched company. Targeted leadership and training of employees convey the applicable standards, methods and rules (see also *induced culture change* Fig. 1). Furthermore, the digitalized structures offer assistance in finding one's way around the storage structures and in minimizing resp. eliminating errors. Both, the orientation system and the ability to adapt and innovate, do not, however, provide reliable security if the interaction partners come from different departments within the company or from other companies, or generally be described as external to the company. In general, there is the possibility of critical interaction situations in which unexpected behavior or unexpected reactions occur on the part of the interaction partners since they cannot derive these from their usual orientation systems.

With respect to the introductory quotation from Confucius [5], it can therefore be assumed that the distinguishing habits are to a large extent traceable to the different cultures. The extent to which culture or cultural heterogeneity influences teamwork

[4]"The set of fundamental beliefs held in common by a group, which is typical of the group as a whole. This set of fundamental beliefs influences the perception, thinking, acting and feeling of the group members and can also manifest itself in their actions and artifacts. The fundamental beliefs are no longer consciously held, they have emerged from the experience of the group and have developed further through the experience of the group, i.e. they have been learned and are passed on to new group members" [32, p. 42].

in digitalized work contexts and the figurations that crystallize in the process will be clarified in the following.

4 Placement of the Research in the Organizational Discourse

Considering the epistemological interest of the article, two lines of discourse appear to be significant. One is the research on (intercultural) cooperation in organizational teams, the other the transfer of figuration-sociological approaches to the workforce in an organization.

With regard to the status quo of the current research discourse on (intercultural) cooperation, numerous studies deal with the effects of diversity on team performance or cooperation within the company. However, the results can sometimes vary depending on what study is consulted. Klöppel [21] and van Knippenberg and Schippers [19], for example, conclude that the advantages and disadvantages in mixed teams cancel each other out and that the probability of success depends on diversity management. Other authors point out the improved possibilities through greater diversity. Cox and Blake [6], for example, elaborate the "value in diversity" hypothesis, from which they derive that diversity has an added value for corporate processes (ibid.). The cultural heterogeneity of the workforce, especially for international companies, is an aspect from which they hope to gain competitive advantages. This hope is guided by the assumption that cultural differences can trigger processes of reflection and learning. As a result, these teams are in many cases more creative, innovative and adaptable [37].

> The findings of our study, therefore, suggest that diversity in teams can potentially provide organizations with competitive advantages if they consider these results in determining the composition of teams while discarding a simple, myopic understanding that team diversity has a uniform effect on team outcomes [17, p. 1009].

Although the hypothesis has been confirmed several times, some research studies show contradictory findings. The teams with homogeneous workforces produced better results at different levels than the heterogeneous teams [17, 18]. One reason for these results is the increased conflict potential—e.g. due to different ages, different cultures, gender differences or language differences—of teams with a diverse workforce [24]. Consequently, different factors influence the success of multicultural teams. The task, the requirement profiles and qualifications of the members, the leadership of multicultural teams, and the working atmosphere and organizational culture play a decisive role [18, 24].

The research outlined so far agrees that the challenges of internal and external cultural diversity are coming to the fore because of advancing globalization, digitalization and the increasing cultural diversity in the organizational workforce. Moosmüller [26], in his contribution to the "Global Organization Man", accentuates the necessity of rethinking for globally operating organizations (ibid.). His focus is on highly qualified white-collar workers and the book "The Organization Man" by Whyte [42]. The "organization man" described by Whyte would have had its day [26]. The "global organization man", which embodies an ideal type of employee who is highly qualified, globally oriented, flexible and interculturally competent, replaces him (ibid.). He concludes that subjectively perceived contradictions are only intensified by facilities such as gated communities[5] and special housing facilities for expatriates. The affected people do not learn about the everyday culture and the prevailing living conditions, but they are forced to pretend this knowledge in order to fulfil expectations. This circumstance leads to the conception of intercultural illusory knowledge (ibid.). Finally, if several globally active employees are brought into the same dilemma, serious damage can be caused by this intercultural pseudo-knowledge at both individual and organizational level. The focus of this chapter is on the highly qualified employees of an organization. Although the present study focuses on low-skilled employees, the work of Moosmüller [26] provides interesting impulses for subsequent research.

Furthermore, Schmidt [33] examines the social relationships in the everyday working life of people of different origin in relation to the functional mechanisms of social integration in companies. In this respect, he elaborates the form of "pragmatic cooperation" (ibid., p. 481). The studies also show a potential for conflict—for example due to the preferences of German employees (ibid.). According to the results, these conflicts gradually dissolve "in the everyday life of cooperation in the work process" (ibid). Even though the companies studied do not have a special diversity management system, the social integration of the employees seems to be positive. However, the normative guidelines are conveyed through a kind of "company universalism" (ibid., p. 482), which is not designed to recognize and emphasize ethnic diversity. In this way, the companies run the risk of institutionalizing the differences of multiculturalism and thus preventing assimilation (ibid.).

The usefulness of applying figuration concepts to the organizational context is demonstrated by Becke [2] in his study on figuration between an organization and external actors, by [3] in their research on the organization itself as figuration, and by Ernst [11] in his observation of figurations within teams in an organization. In these cases, the organizations represent closely related levels of investigation. Rybnikova and Cardone [31] transfer the *established outsider figuration* according to Elias and Scotson [9] to the organizational context and, in the process, examine the temporary employment relationships in a case study from the automotive supplier industry. Among other things, the meaningfulness of transferring figuration-sociological theories to concepts of organizational research clarifies here [31, p. 146]. Furthermore,

[5]The aim of this concept is to provide travelers all over the world with similar living conditions and thus convey a feeling of "home-away-from-home" [26].

the relevance of not regarding the marginal as well as the core workforce as separate, independent actors are emphasized. Rather, they are they are managed as dependent and closely interrelated actors. They constitute their positions of power in interdependent dependence on each other (ibid.). The study by Rybnikova and Cardone [31] provides interesting results and approaches for fundamental considerations on the present research. Nevertheless, multicultural differences or intercultural cooperation are not taken into account. The following research made an attempt to close this gap as well as include and shed light on cultural differences.

5 Field Access and Data Collection

This study deals with the application of everyday practice in the world of employment. In order to do justice to this claim, both participatory observations and qualitative, partially standardized interviews are used to collect empirical data. For the present study, the teams were observed on five working days and one Saturday for the duration of one shift. The observation protocols serve as a documentation of the observations and the interviews conducted in this context. The focus of the observations was on the three teams A, B, C. However, it was possible to gain additional insights into other teams by participating in daily work. This temporal arrangement was chosen as this approach creates a basis of trust and helps to decisively select the interviewees. The interviews took place with a time delay (about one week) following the observations. This temporal arrangement seems to be reasonable, because during the observations, this approach creates already a basis of trust, and it helps to make a decided selection of the interviewees. Qualitative, semi-standardized interviews were used in the present study [16] and are considered a typical form of a qualitative methodological structure in which more or less openly formulated questions are brought into the interview situation in the form of a guideline, to which the interviewee is supposed to respond freely [12]. This form of interview design is particularly suitable for the planned survey, as it offers a diverse scope in the formulation of questions and different demand strategies. Moreover, despite the orientation towards an interview guideline, the sequence of questions to be asked is variable [16]. As a result, in addition to the comparison between interviews, restrictions due to the interview procedure in the intended topics are minimized [35]. In addition, expert interviews with team leaders were conducted.

For the present study, the storage facility and logistics center were considered. In particular, three teams and the corresponding managers are the focus of this research. In this context, a total of eleven interviews were conducted with employees and the team leaders. In addition, observations where performed on six days. This study is thus based on interviews with team members, who represent blue-collar workers [27], and with experts as well as on participatory observations of the cooperation of multicultural teams (means employees with heterogeneous national and/or ethic origin) in everyday working life.

6 Unification of Differences?

In the following, it is essential to illustrate and describe the theory anchored in the empirical data by means of the combination of key and core categories (Fig. 1), as well as to show the location of the identified figuration and the influence of culture in digitalized work contexts. To that end, the model is presented with reference to the core categories *investment in employees*, *permeability*, *storage culture* and *self and third-party construction.*

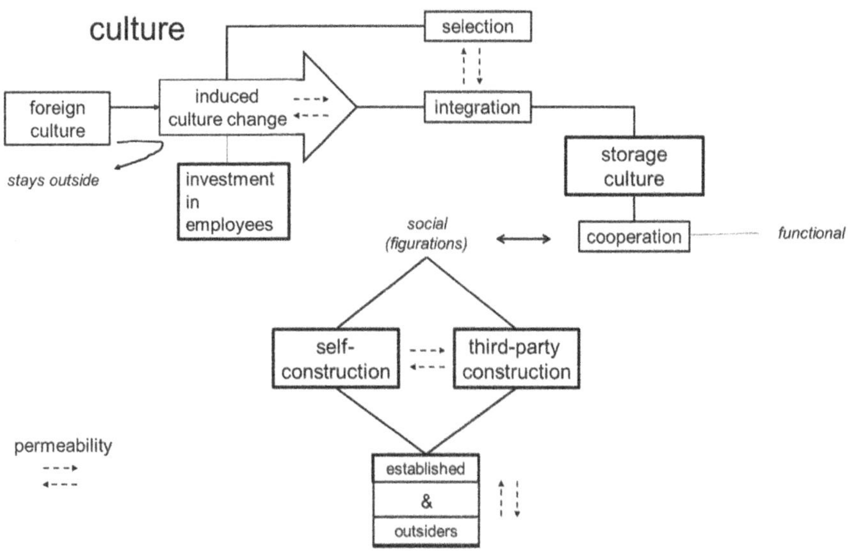

Fig. 1 Model of the unified storage culture (own figure)

An induced cultural change modifies the existing foreign cultures by investing in the employees through leadership, control and by unifying digitalized structures. It must be emphasized that they are not eliminated or rejected. Moreover, the induced culture change is not permanently modifying them outside the storage context.

The definition of culture is largely transferable to the phenomenon of *storage culture*. In this way, a separate *storage culture* is forming in which the various foreign cultures play no or a very subordinate role. Nevertheless, the various cultures influence the *storage culture* in its constitution. In order to achieve the goal that the different cultures play a subordinate role in the *storage culture*, a management adapted to the cultural differences is necessary. The cultures must consequently play a role in order to then play no or a small role.

Investment in Employees and Permeability

When leaving the storage facility, whether during breaks or at the end of a working shift, the individuals fall back into their old, native patterns of behavior and classification. The employees experience this permeability several times a day. The term permeability in the conceived theory stands for the stringent transparency of the process. This is where the dynamic character of the present model becomes apparent: The process is not only possible in one direction, there is always the possibility to return to the previous state.

Accordingly, the investment in employees represents an important aspect for the modification of the foreign culture into the *storage culture*. In this case, it is a form of management and control over the team members. Ideally, they get to know what they are supposed to convey as well as the values and norms through explanation in order to create understanding and to secure them in their work:

> Then, however, their security comes into play—that is a great need for me [...] you just have to explain to people why they are doing something and then working is also much easier afterwards (employee, Team C, line 42).

The type of leadership is important in order to achieve the desired result of an integrated and accepted *storage culture*. At this point, a discrepancy between self-construction and external construction becomes apparent. The team members do not necessarily have the feeling to experience an investment:

> I said: 'So where's the problem? This is my workplace. So why is there a problem. You don't work here.' 'Well, cause I'm the boss' That's his answer. I said 'Ohhh now I understand. You're one of those kind of team leaders. 'No matter what I do 'I'm the boss' I say ahhh ok. Great. Great team leaders' (employee, Team B, line 46).

The team leaders see this kind of leadership as an investment: "Well, we have a lot of patience. If you could see what we invest in people, how we respond to personal events and so on" (team leader, Team B, line 32).

Managing a multicultural team is a challenge for the management despite the development of a unified *storage culture*. At the same time, this process is an important milestone, as it initiates the induced cultural change. It "has now settled down a little bit, a year and a half or so. In the meantime, everyone has his or her own areas under control. You can already see an increase in quality and efficiency" (team leader, Team B, line 28). To further consolidate these goals, the development of intercultural competence is desirable. Digitalized structures support this development through simplification by unification of e.g. communication channels. At this point it is also important to note that the introduction and integration into new digitalized structures requires adapted and participative management behavior in order to integrate all employees into the *storage culture*. Furthermore, important aspects of management and control are explaining, understanding and the change of perspective of the employees. The individual support and leadership of individual employees is particularly important in view of the multicultural workforce and the associated differences in understanding leadership, hierarchies and security regulations; this is how the change towards a *storage culture* is processed. It must be taken into account that this process with the goal of a uniform *storage culture* is a time-consuming process. The actors with their diverse cultures contribute to the *storage culture* or influence it. The multicultural teams are also an opportunity to gain knowledge. It tries to feed in the advantages from each culture. Two possibilities result from the investment in the employees: selection of members or adaptation and integration.

Storage Culture

During the selection process, employees are either not hired, dismissed or the issue shifts to another area: For example, in the sales area, multiculturalism counts as a company strength, as customers can be served in their native language. However, the context of the storage facility and thus the *storage culture* is left behind—accordingly, other orientation systems take effect. The adaptation and integration of employees leads to the adoption of the orientation and rule system applicable in the *storage culture*. This ideal of the *storage culture*, which is in development and already partially in place, bases on a strong team orientation of the team members and harmonious cooperation. This results in an orderly process flow that facilitates cooperation. In this regard, these processes and the striving to achieve goals represent a functional component. The need for harmony and the great responsibility of the individual (if the ideas of the *storage culture* are effective), among other things, for the maintenance of the *storage culture*, connect the functional component with the social component and the figurations of *the established* and *outsiders* that arise here. Accordingly, the concept of *storage culture* as a system of orientation is only permanent in the storage facility. As soon as the employees leave the storage, the prevailing culture no longer is an orientation system. This circumstance results from the strict permeability of the model and the lack of investment in staff outside the storage facility. Thus, when leaving the storage facility, not all the premises necessary for the existence of the *storage culture* are fulfilled. Consequently, several requirements have to be met to give validity to *the storage culture* as a socially constructed, complex and dynamic orientation system: On the one hand, the actors are in the working context and the frame of reference of the locality—the storage—is fulfilled. On the other hand, there is a necessary investment in the employees in form of leadership and control. An established *storage culture* can certainly contribute to the goals of harmonious cooperation, team stability, orderly process flow, and thus to functional cooperation. "But I think harmony goes beyond cultures, that's true, isn't it?" (employee, Team C, line 88).

Self and Third-Party Construction

The *established outsider figuration* by self- and third-party construction. The team leaders represent *the established* and the team members *the outsiders*. The individual attributing the corresponding characteristics to himself, defines the self-construction whereas the external construction contains the characteristics that attributes to the individual or the team by others. These constructions have a decisive influence on the figuration: If, for example, a self-construction is strong enough to resist the image of the others for a certain period, it is possible that it will be adapted as an image of the others by the other actors and change accordingly. There is also the possibility that the self-image bends to the image of the other. The discrepancy between self-construction and external construction results in a lack of team stability, a high workload and uncertainty. If the self-image and the image of the other were harmonious, this would possibly result in greater team stability and a lower workload. In this figuration, however, a special feature constitutes: There is not only a management level and a worker level, but also an intermediate level on which there are employees with such a strong positive self-image that they see themselves in an unofficial leadership role. The employees and, to a limited extent, even the management bodies accept this role. This intermediate level reflects that of the unofficially established or outsiders. Thus, the permeability of the process steps and the dynamic flow of power are reflected here. *The outsiders* can become established; conversely, *the established* can fall into the role of outsiders. The latter ends, however, as in the observed case studies, in leaving the company. The hierarchical structures as well as the monitoring and control of the employees strongly influence the relationship networks. The hierarchical structure becomes evident here. Only the established ones monitor and control *the outsiders*. Thus, *the established* have the possibility to influence *the outsiders*, but *the outsiders* do not have the possibility to supervise the established. Not least because of this fact, the existing power gap in the figuration becomes evident. This in turn affects cooperation through the created hierarchical structures. The figuration is thus dependent on the cooperation, but at the same time, the cooperation influences the figuration. Regarding the identified figuration, the permeability becomes clear at this point: There is the possibility of a transition between outsiders and the established, the result of which is a group of unofficially established people. In addition, there is also permeability in the other direction—from the established group to *the outsiders*—but this usually leads to leaving the company and thus joining the group of former members (Fig. 2).

Through the interaction of all core categories, a form of *storage culture* emerges; it influences the cooperation in organizational teams and contributes to a unification of the differences. It is important, however, that despite—or precisely because of—the *storage culture* (or the goal of the *storage culture*), the multicultural workforce requires attentive and individual leadership. The storage facility is therefore an ideal type, which is predominantly found in the observed company, with few exceptions.

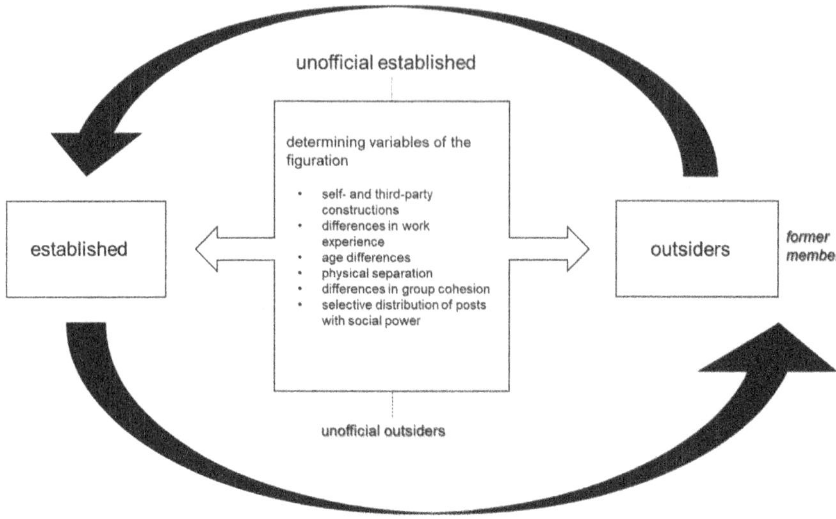

Fig. 2 Determining variables of the figuration (own representation based on [31])

An elementary function is assigned to the digitalized structures in the company. For example, the employees are enabled to carry out self-control by the scanners used. "I first look at something in the computer, in the scanner or in the computer, what I should do and more work. And then I do" (employee, Team A, line 14). The instructions and work orders are uniformly formulated—communication with third parties is not necessary. If an employee takes a wrong article from the shelf, the scanner will provide an auditory warning and the digital system blocks further processing of the process. At this point, the extent to which the digitalized structures contribute to the unification of differences becomes apparent.

7 Discussion

Prior to reflecting and summarizing the research process in the following section, the results of the present study need to be discussed. The central concerns of the research were pointing out the figurations that arise in the course of cooperation in digitalized work contexts in organizational teams and in clarifying the question of whether culture exerts an influence on team cooperation and if so to what extent.

Attempting to transfer the *established outsider figuration* to the context of the organization and to shed light on cooperation in multicultural teams yielded interesting results, although these differ in part from those by Elias and Scotson [9]. In summary, the characteristics of the determinants of an *established outsider figuration* reveal a model based on the premise of two interdependent groups with a power imbalance. The trigger can be a competitive situation, which expresses, for example, in uncertainty or in the fear that one's own social status is at risk. This results in the rejection of *the outsiders* by the established. The group of *the established* has the possibility to build up a strong group cohesion, while *the outsiders* do not develop it:

> Its greater cohesion enables such a group to reserve social positions with a high power potential of a different type for its members, thus in turn reinforcing its cohesion, and to exclude from them members of other groups which is essentially what one means when one speaks of an established outsider figuration [9, p. 19].

Transferred to the figuration compiled in the course of the present research, the *established outsider figuration* appears—albeit in a special form—between the team leaders (*established*) and the team members (*outsiders*). The cohesion in the group of team leaders is higher than in the group of team members, not least because of the high fluctuation in the group of team members. The team members do not have the possibility to build up a firm relationship among each other due to the constant change in the team. At the same time, the team leaders sometimes prevent personal contact. Moreover, the *established outsider figuration* differs from that by Elias and Scotson [9]. Here an intermediate dimension of the unofficially established or outsiders becomes apparent. Although officially they belong to the respective group, they are at the transition to the opposite group (Fig. 2). The permeability of the process, which runs through the entire model, is also visible here.

In addition to these results, the *storage culture* is a phenomenon that is occasionally not necessarily described as a desirable one. In theory, the aim is to create an intercultural encounter which is not forcing the foreign culture into the own orientation system. However, it seems to make sense to establish a uniform *storage culture* in the observed company. In view of the results of the study and the research field, it is also noticeable that even though the challenges of intercultural cooperation addressed in the theory do exist, the synergies do not inevitably become effective in this case study. Thus, an individual management geared to the multicultural workforce seems to be inevitable, although the solution is to adapt and integrate into the uniform *storage culture*. Previous research primarily deals with the cooperation of multicultural teams in the field of white-collar workers. The results obtained in this respect cannot be confirmed in the present research. Rather, the different cultures play a subordinate role during work in the storage facility: "You have a job to do, it has to be done, over! Finished! You don't need any, I don't know, the Yank can do it better, the Syrian can do it better, you don't have that in the storage facility. Everyone has to do his work" (team leader, Team B, line 77). Also, considering the introductory statement of Confucius, the uniform habits of storage work in the digitalized work contexts lead to a unification of the differences.

8 Conclusion and Outlook

The research-guiding question of the formation of figurations in digitalized work contexts in organizational teams and the influence of culture on cooperation was answered based on the model of a *unified storage culture* derived from the empirical data. During this process, an *established outsider figuration*, which the team leaders and the team members define, crystallizes. The leadership and control of multicultural teams creates different challenges, but as an induced cultural change it ideally leads to a *unified storage culture*. Digitalized structures in the company, which simplify the unification of the differences, support this development.

Finally, the basic assumptions of figuration sociology are reflected in the research process. The *established outsider figuration* results from the interplay of interdependent relationships between team leaders and team members. In addition, the relationships and dependencies between them drive the transformation from foreign cultures to *storage culture* forward. Despite the relatively short research period, the long-term nature of the social processes became clear, and the model and its concepts were observed in their 'becoming'. The establishment of the storage culture was not an intentionally planned process, and it was also not an aimless one. Likewise, the interdependence and relative autonomy of the actors are expressed in the figuration described, in the *storage culture* and its effects; consequently, they are free in their actions to a limited extent and are only interwoven with each other through figurations.

The strengths of the research work lie in the precisely defined research process, through which traceability is guaranteed. In addition, a model based on the empirical data was developed, which provides initial starting points to shed light on aspects surrounding the conception of a *storage culture* in the local company and the formation of figurations in this context. A certainly relevant limitation of the present study is the survey and evaluation procedure of the *grounded theory methodology*. Although this procedure has advantages that have been described many times, these should be critically reflected, especially regarding possible losses of data material during the reduction by forming categories. A further limiting aspect in this context is the question of repeatability, which is indispensable in quantitative research [36]. Although the results could be tested for causality, it should be noted that for social phenomena an exact repeatability of the study with identical results is not possible. In view of the processual nature of social reality, it is not possible to create identical initial conditions for the renewed study [36].

Furthermore, it is worth noting that the results of this research work originate from a situation-related case study and the model does not claim to be universally transferable. Nevertheless, the assumptions provide an adequate framework for deriving further research questions. For example, research into team-building processes, or the integration of diversity management in order to implement the induced cultural change in a more targeted and effective manner would be relevant and current topics for subsequent research questions. Additionally, the question of the influence of other areas of the company on the *storage culture* arises. Moreover, there is the possibility to test the research for its transferability by trying to apply it to other companies. For example, in terms of digitalization, the figuration between white and blue-collar workers would be interesting. Looking at the current situation (COVID-19 Pandemic), it becomes obvious that especially blue-collar workers are not benefiting from digitalization to the full extent. This is particularly evident in the (pandemic prevention) measures for working in home offices see for further information Pauly and Holdampf-Wendel [29]. The employees must work on site.

Finally, it is worth mentioning that—not least because of the globalization and digitalization—there are hardly any alternatives to multicultural teams in organizations. Therefore, the question is not whether not whether these are in use, but how they deal and the challenges they pose can be dealt with. Targeted guidance and support through digitalization play a central role in the researched company in order to achieve that the differences in the unified *storage culture* play a subordinate role. This chapter provides initial approaches and impulses for thinking in order to shed light on this topic and to derive possible recommendations for action in entrepreneurial practice.

References

1. Baur N, Ernst S (2011) Towards a process-oriented methodology: modern social science research methods and Norbert Elias's figurational sociology. Sociol Rev 59(1):117–139
2. Becke G (2002) Wandel betrieblicher Rationalisierungsmuster durch Mitarbeiterbeteiligung. Eine figurationssoziologische Fallstudie aus dem Dienstleistungsbereich. Campus, Frankfurt
3. Becke G, Wehl R (2016) Generationengerechtigkeit als Schlüssel für betriebliche Sozialinnovationen. In: Becke G, Bleses P, Frerichs F, Goldmann M, Hinding B, Schweer M (eds) Zusammen–Arbeit–Gestalten. Soziale Innovationen in sozialen und gesundheitsbezogenen Dienstleistungen (pp 223–242). Springer VS, Wiesbaden
4. Bourdieu P (1977) Outline of a theory of practice Cambridge studies in social and cultural anthropology. Cambridge University Press, Cambridge
5. Confucius (1979) The Analects, tr. D.C. Lao. Penguin Books, Harmondsworth
6. Cox TH, Blake S (1991) Managing cultural diversity: implications for organizational competitiveness. Executive 5(3):45–56
7. Elias N (2000) The civilizing process. sociogenetic and psychogenetic investigations. Blackwell Publishing, Malden
9. Elias N, Scotson JL (1994) The established and the outsiders. Sage, London
10. Elias N (2006) Figuration, sozialer Prozeße und Zivilisation: Grundbegriffe der Soziologie. Aufsätze und andere Schriften III. In: Blomert R, Hammer H, Heilbron A, Treibel N, Wilterdink (eds) Norbert Elias Gesammelte Schriften (vol 16) (pp 100–117). Frankfurt a. M.: Suhrkamp
11. Ernst S (2010) Prozessorientierte Methoden der Arbeits-und Organisationsforschung. Eine Einführung. Springer VS, Wiesbaden
12. Flick U (2007) Qualitative Sozialforschung. Eine Einführung. Rowohlt, Hamburg
13. Frerichs M (2014) Innovationsprozesse und organisationaler Wandel in der Automobilindustrie: eine prozesssoziologische Analyse betrieblicher Machtproben. Springer VS, Wiesbaden
14. Hirsch-Kreinsen H (2016) Digitalisation and Low-Skilled Work. WISO Diskurs 19/2016. Friedrich-Ebert-Stiftung, Bonn
15. Hofstede G (2001) Culture's consequences, 2nd edn. Sage, London
16. Hopf C (2015) Qualitative Interviews—ein Überblick. In: Flick U, von Kardorff E, Steinke I (eds) Qualitative Sozialforschung. Ein Handbuch (pp 349–360). Rowohlt, Reinbeck

17. Horwitz SK, Horwitz IB (2007) The effects of team diversity on team outcomes: a metaanalytic review of team demography. J Manage 33:987–1015
18. van Knippenberg D, de Dreu CK, Homan AC (2004) Work group diversity and group performance: an integrative model and research agenda. J Appl Psychol 89:1008–1022
19. van Knippenberg D, Schippers MC (2007) Work group diversity. Annu Rev Psychol 58:515–541
20. Kroeber A, Kluckhohn C (1952) Culture: a critical review of concepts and definitions. Cambridge: The–Peabody–Museum of American Archeology.
21. Köppel P (2007) Konflikte und Synergien in multikulturellen Teams. Deutscher Universitätsverlag, Wiesbaden
22. Maletzke G (1996) Interkulturelle Kommunikation. Zur Interaktion zwischen Menschen verschiedener Kulturen. Westdeutscher Verlag, Opladen
23. Marquardt MJ, Horvath L (2001) Global teams—how top multinationals span boundaries and cultures with high-speed teamwork. Palo Alto, California
24. Mensi-Klarbach H (2011) Interkulturalität im Arbeitskontext – Zwischen Konflikt und Chance. In: Mallich K, Gutiérrez-Lobos K (eds) Mehr Chancen durch Vielfalt! – Diversity in der Personalentwicklung. Medizinische Universität Wien, Wien, pp 52–67
25. Moosmüller A (2009) Kulturelle Risiken der internationalen Unternehmenstätigkeit. In: Kühlmann TM (ed) Internationales Risikomanagement: Auslandserfolg durch grenzüberschreitende Netzwerke (pp 61–81). Oldenbourg, München
26. Moosmüller A (2014) Global organization man. In: Jammal E (ed) Kultur und Interkulturalität. Interdisziplinäre Zugänge (pp 169–184). Springer VS, Wiesbaden
27. Oberbeck H (2017) Angestelltensoziologie. In: Hirsch-Kreinsen H, Minssen H (eds) Lexikon der Arbeits- und Industriesoziologie (2nd edn). Nomos, Baden-Baden
28. Orwell G (2003) Animal farm and 1984. Harcourt, Orlando
29. Pauly B, Holdampf-Wendel A (2020) Corona-Pandemie, Arbeit im Homeoffice nimmt deutlich zu. https://www.bitkom.org/Presse/Presseinformation/Corona-Pandemie-Arbeit-im-Homeofficenimmt-deutlich-zu. Accessed on 30 May 2020
30. Podsiadlowski A (2002) Multikulturelle Arbeitsgruppen in Unternehmen. Bedingungen für erfolgreiche Zusammenarbeit am Beispiel deutscher Unternehmen in Südostasien. Waxmann, Münster
31. Rybnikova I, Cardone P (2018) Weder Etablierte noch Außenseiter. Figurationssoziologische Betrachtung der Leiharbeit [en. Neither established nor outsiders: considering temporary agency work from the perspective of figurational sociology]. Managementforschung 28:117–149
32. Sackmann S (2017) Unternehmenskultur: Erkennen – Entwickeln – Verändern. Erfolgreich durch kulturbewusstes Management, 2nd edn. Gabler, Wiesbaden
33. Schmidt W (2006) Pragmatische Zusammenarbeit. Kollegialität und Differenz bei Beschäftigten deutscher und ausländischer Herkunft in Industriebetrieben. Zeitschrift für Soziologie 35(6):465–484
34. Schönhuth M (2017) Das Kulturglossar. Ein Vademecum durch den Kulturdschungel für Interkulturalisten [Glossary culture and development: A vade mecum through the culture jungle]. Trier, Germany: Trierer Materialien zur Ethnologie.] [last downloaded 2020.05.30]
35. Strübing J (2013) Qualitative Sozialforschung: eine komprimierte Einführung für Studierende. Oldenbourg, München
36. Strübing J (2014) Grounded Theory. Zur sozialtheoretischen und epistemologischen Fundierung eines pragmatistischen Forschungsstils, 3rd edn. Springer VS, Wiesbaden
37. Süß S (2007) Diversity-Management als integratives Konzept zum Umgang mit kultureller Heterogenität im internationalen Unternehmen. In: Oesterle M-J (ed) Internationales management im Umbruch. Gabler, Wiesbaden, pp 291–319
38. Thomas A (1993a) Kulturvergleichende Psychologie. Hogrefe, Göttingen
39. Thomas DA (1993b) Racial dynamics in cross-race developmental relationships. Adm Sci Q 38(2):169–194

40. Thomas A (2010) Theoretical basis of intercultural communication and cooperation In: Thomas A, Kinast EU, Schroll-Machl S (eds) Handbook of intercultural communication and cooperation. Vol. 1: basics and areas of application, 2nd rev. ed. Göttingen: Vandenhoeck & Ruprecht
41. Treibel A (2009) Figurationsund Prozesstheorie. In: Kneer G, Schroer M (eds) Handbuch soziologische Theorien (pp 133–160). Springer VS, Wiesbaden
42. Whyte WH (1956) The organization man. University of Pennsylvania Press, Philadelphia
43. Willems H (2010) Figurationssoziologie und Netzwerkansätze. In: Stegbauer C, Häußling R (eds) Handbuch Netzwerkforschung (pp 255–268). Springer VS, Wiesbaden
44. Witte EH (1994) Lehrbuch Sozialpsychologie, 2nd edn. Beltz, Weinheim

Mona-Maria Bardmann is research assistant at the Institute of Education, Work, and Society at the University of Hohenheim in Germany. She studied sociology, philosophy, and media and cultural sociology at Trier University. Her research interests are in the field of sociology of work, human cooperation in organizations and sociology of culture.

Evaluation

Efficiency Measurement in Digitalized Work Systems of Transport Logistics

Dominic Loske and Matthias Klumpp

Abstract Digitalization is a major trend and challenge in most industries and sectors of societies. Still, quantitative insights regarding the impacts of digitalization are missing. This chapter is reporting a first approach using Data Envelopment Analysis (DEA) for measuring efficiency results of digitalization steps in a retail logistics context. Aspiring to quantify the performance of professional truck drivers during a digital turnover related to mobile devices, we evaluate truck loading processes. As inputs we use loading time and costs. Outputs are load factor of units, invoice charged to shops, and the value of the damages during truck loading. The findings indicate that a change in the level of digitalization entails a loss of the efficiency level in the first instance, which can be compensated and even surpassed later. When applying linear regression analysis, we prove a low statistical linear relationship of age and efficiency plus a strong statistical linear relationship of employer size and efficiency as well as period of employment and efficiency, always regarding the changing levels of digitalization in the working system of professional truck drivers. For practitioners in retail logistics, we derive the importance of employee retention programs for human resource management, along with a positive working environment provided for truck drivers to reduce fluctuation effects. Furthermore, we advise designing software for truck drivers as commonplace as possible and in the style of widespread smartphone software user interfaces.

D. Loske (✉)
Institute for Logistics and Service Management, FOM University of Applied Sciences, Leimkugelstr. 6, 45141 Essen, Germany
e-mail: dominic.loske@fom-net.de

M. Klumpp
Chair for Production and Logistics, Georg-August-University of Göttingen, Platz der Göttinger Sieben 3, 37073 Göttingen, Germany

Fraunhofer IML, Dortmund, Germany

FOM University of Applied Sciences, Essen, Germany

M. Klumpp and C. Ruiner (eds.), *Digital Supply Chains and the Human Factor*, Lecture Notes in Logistics,
https://doi.org/10.1007/978-3-030-58430-6_9

1 Introduction

The current scientific discussion, focusing on the development and application potentials of information- and communication-technology, is reflected in keywords such as Cyber-Physical Systems (CPS)[1] regarding the production area, Internet of Things (IoT)[2] in the logistics and transportation area and Artificial Intelligence (AI)[3] within the field of informatics. Based on a sector-specific perspective and the related revolutionary development, the innovation advances are furthermore discussed under buzzwords like Industry 4.0,[4] Logistics 4.0,[5] or Retail 4.0.[6] Referring to the digital transformation described within these concepts, the current research discussion postulates almost unanimously a development boost that is characterized by a new and, in its extent, unknown evolution, with non-foreseeable application potentials [12]. The social and economic consequences can be described as disruptive and affect logistics as a scientific discipline, activity, and industry [58, 99]. Yet, scientists and practitioners know little about what exactly is behind these buzzwords and how the emerging changes of working systems will transform work tasks or working relationships—and especially the question of efficiency improvements is without widespread empirical answer yet. Being able to derive interesting research questions requires to identify a research gap regarding the impact of digitalization on the human workforce within logistical work systems.

1.1 Research Gap Regarding the Impact of Digitalization

To identify an existing research gap, the development boost mentioned in Sect. 1 was analyzed by a systematic bibliometric analysis based on Elsevier's abstract and citation database Scopus. Therefore, the focus was on publications dealing with the impact of digitalization between 2008 and 2018 within the document types of textbooks, conference papers, and journal articles. Furthermore, publications related

---

[1]See Broy and Geisberger [21] and for the basic architecture of CPS systems, see [64, 65, 76, 82].

[2]The basic idea of this concept is the existence of several objects that are able to interact and communicate with each other, see [11]. For basic literature regarding the IoT [3, 41, 75, 107].

[3]The umbrella concept AI contains an intelligence created by human skills and was first-ly named by McCarthy et al. [74]. A wide range of existing definitions can be categorized in four main approaches [85] (1) thinking humanly [43], (2) acting humanly [61], (3) thinking rationally [24] and (4) acting rationally [81].

[4]See Kagermann et al. [51, 52], Forschungsunion [37, 38], Forschungsunion and Acatech [39]. Industry 4.0 is also mentioned as "Industrialization of a new type", see [18], as "Second Machine Age" see [22], as "Third Industrial Revolution", see [72, 84] and as "Distributed Capitalism", see [110].

[5]See [40, 50, 68, 78, 87, 94, 103].

[6]The concept drafts of Retail 4.0 are e.g. anticipatory shipping through big data analytics see Lee [63] and consumer's purchase prediction [14, 73].

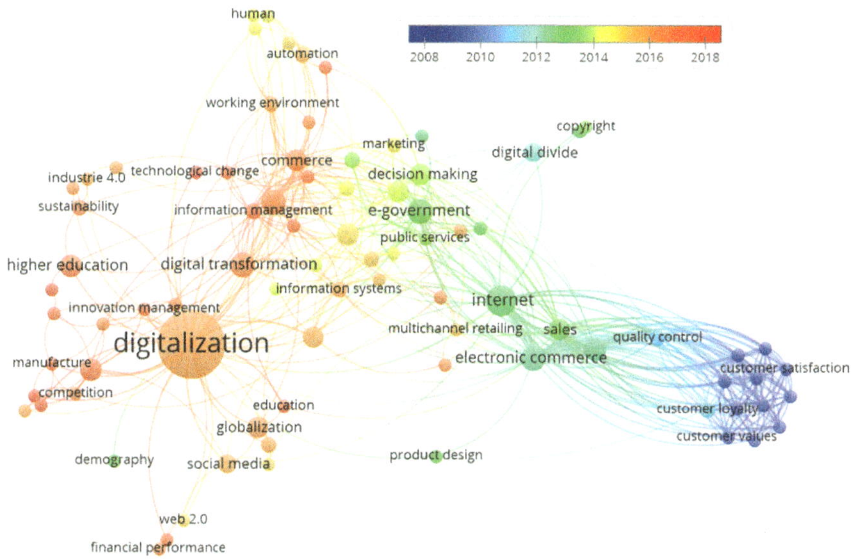

Fig. 1 Keywords map of investigated publications regarding the impact of digitalization

to the sections of social sciences or business/management/accounting were investigated. TITLE-ABS-KEY ("impact" AND "digitalization") AND PUBYEAR > 2008 AND (LIMIT-TO (SUBJAREA, "SOCI") OR LIMIT-TO (SUBJAREA, "BUSI")) was used as search term. To illustrate the 154 resulting publications, VOSviewer, a software tool for constructing and visualizing bibliometric networks, was applied [98]. The bigger a circle, the more frequently the keyword appears in the publication set from Scopus. The keywords are assigned to a cluster based on a computer algorithm. Each cluster has its own color related to the year of publication (Fig. 1).

The results show that the current research discussion concentrates on the investigation of digitalization in the industrial sector (keyword, e.g., Industry 4.0, manufacturing) and, moreover, on the impact on the human workforce and existing working systems (keyword, e.g., human, working environment, education) (Fig. 2). Therefore, the blue-collar workers in the logistics sector are hardly addressed.

The research approaches of the publications are either conceptual frameworks [32, 35, 45, 46, 49, 56] or empirical studies based on qualitative research methods [1, 16, 57, 93, 101, 102, 104]. Consequently, there is a research gap for investigating the impacts of digitalization in logistics based on a quantitative methodology.

1.2 Research Question and Methodology

This contribution aims to develop and apply an efficiency analysis to evaluate the effect of changing levels of digitalization within the working systems of transport

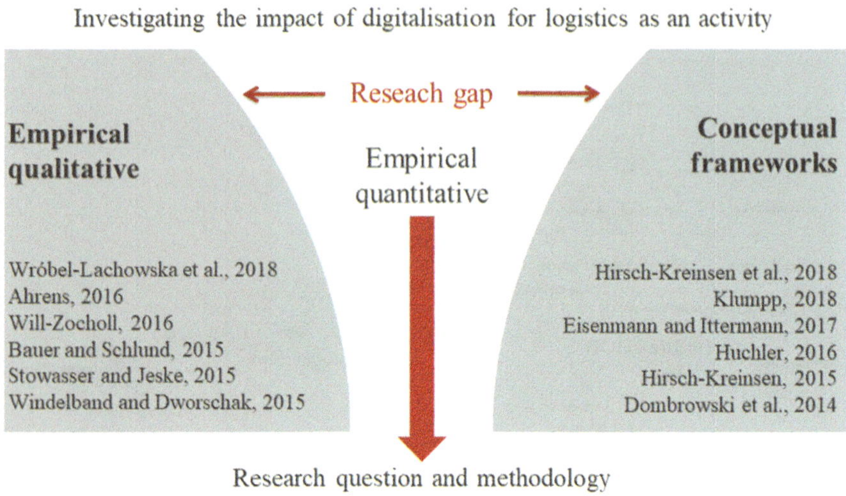

Fig. 2 Research gap regarding the impact of digitalization

logistics. The understanding of the term efficiency and its demarcation to the term effectiveness is based on the contents of ISO 9000: [31] in this contribution. Therefore, efficiency is defined as the "ratio between the results achieved and the resources deployed" and effectiveness as "the extent to which planned activities are achieved and planned outcomes are reached [31, p. 22]." The contribution aspires to answer the following four research questions:

- RQ_1: What are the components of an efficiency analysis aimed at evaluating the effect of changing levels of digitalization within the working systems of professional truck drivers?
- RQ_2: Which level of efficiency exists before the change in the level of digitalization, how does it develop during and after the changeover?
- RQ_3: Does the efficiency analysis determine a gain or a loss of efficiency for the examined working systems?
- RQ_4: What effect do demographic factors such as age, employer size, the period of employment, migration background, and job profile of parents have on the level of efficiency after a change in the level of digitalization?

After this introduction in Sect. 1, relevant theoretical basics regarding the theory of the working system and the theory of efficiency in transformation processes are presented in Sect. 2. In the next chapter, the efficiency analyses for digitalized working systems are elaborated by presenting relevant basics on DEA and by formulating a DEA model using appropriate inputs and outputs in Sect. 3. The presentation of the empirical results in Sect. 4 is the basis for further correlation and regression analyses in Sect. 5. After deriving implications for logistics managers in Sect. 6, the contribution will end with a conclusion and outlook in Sect. 7. Figure 3 summarizes the methodology of this contribution.

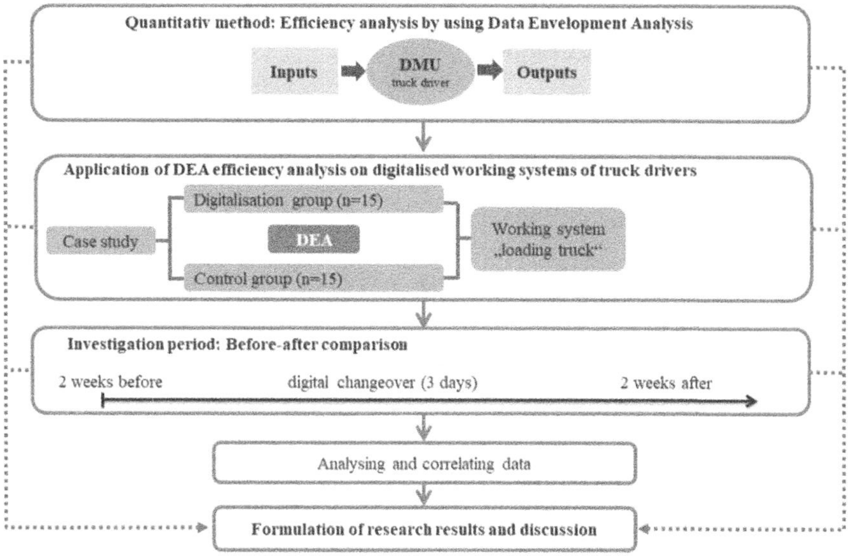

Fig. 3 The methodology of the contribution

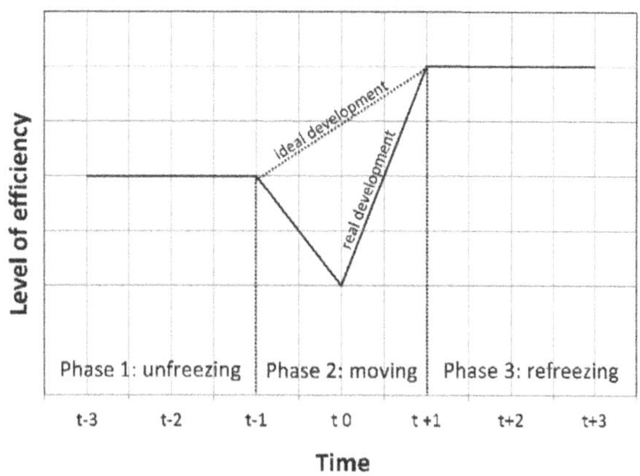

Fig. 4 Efficiency progression in transformation processes

2 Literature Review

To identify the most relevant scientific contributions, two systematic and iterative bibliometric studies are implemented for (1) concretizing the object of research for all case analyses in Sect. 2.1. and for (2) creating a theoretical basis regarding the efficiency in transformation processes for the transportation centered case analyses in Sect. 2.2. For this purpose, Elsevier's abstract and citation database Scopus and Clarivate Analytics, database Web of Science were used. It was possible to prove that there is currently no approach aiming to simulate efficiency levels based on a quantitative empirically determined curve progression for digital transformations and learning progression scenarios.

2.1 *The Theory of the Working System as a Frame of Reference*

The digital transformation examined in this contribution is not related to the retailing company according to a holistic organizational view, but rather to a subsystem where transport-logistical activities are conducted on an operational level. Therefore, the objects of research within the case analyses are work systems, which are defined as:

> A system in which human participants and/or machines perform work (processes and activities) using information, technology, and other resources to produce specific products/services for specific internal and/or external customers [10, p. 75].

Therein the work system framework, as a socio-technical system, plays an important role and is formulated, applied and extended for more than 20 years [4–10, 42, 69]. To accomplish the working task, the human resource impinges on the work object by using work equipment in order to achieve a given work objective [69]. The interaction between humans and the work equipment is necessary and requires a certain level of qualification for successful interaction. Besides the inputs into the working system (e.g., energy, material, and information) and the output factors (e.g., work results), there are exogenous factors influencing the working system and subsystems like the physical and social environment [88]. Within the case analyses related to the transport logistics sector, the truck drivers reach the objective of a complete and digitally recorded truck loading by using mobile handheld scanners. The qualification requirement and a digital transformation scenario affect the track and trace system.

2.2 *Theory of Efficiency in Transformation Processes*

Scientific contributions exploring changes in organizations are often associated with change management [97]. There are theoretical-conceptual approaches that describe organizational changes on a macro level, e.g., as Kotter's 8-step process [60], or at the micro-level, e.g., as Lewin's 3-stage model [66, 67] (Fig. 4).

Lewin was one of the first scientists to study the efficiency and impact of change, especially the reactions of groups facing planned changes [19]. According to Lewin's model, the process of change can be divided into three phases, referred to as unfreezing, moving, and refreezing [66, 67]. In addition, the model indicates individual efficiency levels depending on the current change phase. The bibliometric analysis conducted for this publication did not identify scientific contributions that aim to simulate the presented curve progression by empirical quantitative data when examining digitalization in retail logistics.

3 Efficiency Analysis and Digitalization in Transport Logistics

Contributions dealing with efficiency measurement often distinguish between (1) parametric, (2) semi-parametric, and (3) non-parametric methods, which can be assigned to performance measurement. Therein, (1) parametric methods require the functional relationships of the observed factors a priori, which can be achieved by applying the deterministic method Ordinary-Least-Square (OLS) or stochastic approaches such as Stochastic-Frontier-Analysis (SFA) [2]. The criticism of the parametric methods lies in the a priori assumption, whereby an assumed production function is adapted on empirical data by statistical methods, and all remaining deviations are then statistically determined [53]. In contrast to these approaches, (2) semi- and (3) non-parametric methods do not require a given functional relationship of included factors [23]. In particular, this includes stochastic Data Envelopment Analysis (DEA) and deterministic Data Envelopment Analysis, which was first published in Charnes et al. [23].

Besides the existing methods of efficiency measurement, the aim of this contribution is to measure the digitalization phenomenon. Thereby, four characteristics of the digital transformation justify the application of DEA: First, the impact of digitalization is not predictable and not yet examined from an efficiency-based point of view. Therefore, there is no a priori knowledge about functional relationships of digitalization towards humans. Second, because the digitalization is influencing the human workforce, the theory of work systems is applied as a theoretical framework within the case analyses. As the achievement of work objectives requires input and produces output, a method that enables the integration of several in- and output factors along with the possibility of factor enhancement are postulated. Third, without the existence of a benchmark value for the level of efficiency in digitalization scenarios, the analysis has to compare the performance of the empirical observation among each other. Fourth, as it is unclear if digitalization spawns an immediate or gradual efficiency development, the progress of efficiency in retail logistics is illustrated with an empirical curve progression. Therefore, the results of the analysis have to be comparable among several periods. As DEA does not require a priori information (requirement 1), considers multiple measures (requirement 2) [28, 29], compares solely the empirical observation among each other (requirement 3) [23] and the results are comparable when factors are constant (requirement 4) [28, 29], it is the method of choice aiming to answer RQ_1.

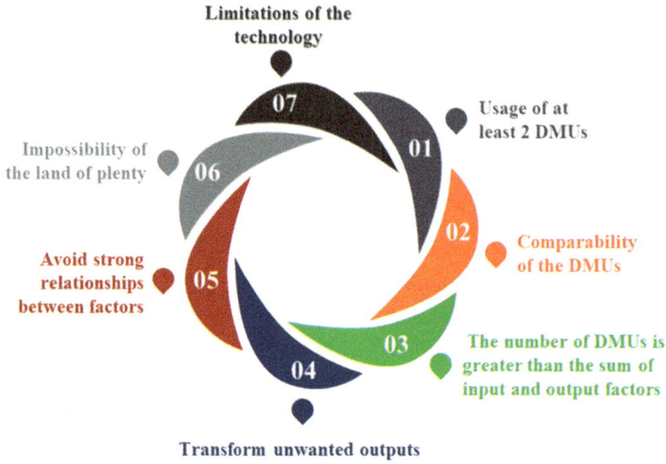

Fig. 5 Application requirements of DEA

3.1 Data Envelopment Analysis for Efficiency Measurement

Based on Koopmans' work on activity analysis as well as Debreu and Farrell's work on radial efficiency measurement [30, 36, 59], Charnes et al. developed a method for determining the efficiency of so-called decision-making units (DMU) for multiple inputs and outputs [23]. Initially presented as an evaluation method for non-profit organizations,[7] scientists recognized the interdisciplinary application potentials DEA offers.[8] DMUs can be organizations or organizational units such as banks [25, 48], educational institutions [55, 62, 105], airlines [26, 79], or persons [13, 33, 34, 54, 91, 108], branded products [17], and nations [70, 71]. The relationship between inputs and outputs is determined by non-parametric methods, and are considered as a black box [86]. The efficiency of a certain technology is not prescribed by a production function. Rather, a minimally enveloping function course is approximated overall empirically observed values, whereby at least one value of the observations must lie on exactly this function, the so-called best-practice line [92]. The advantage of this black box assumption is that no a priori assumptions about the design of the production function must be made [28]. Figure 5 summarizes all application requirements of DEA.

The ability to find virtual reference units on a best-practice straight line, assuming constant returns to scale (CRS) as well as an input or output orientation, is just one of two major optimization steps in the CCR model. In addition, DEA is able to weight the factors discriminating their value model-endogenously [23]. The weights

---

[7]In the original version "...for use in evaluating activities of not-for-profit entities participating in public programs"; Charnes et al. [23, p. 429].

[8]The bibliography of Tavares lists with 3.203 analyzed publications of 2.152 authors from 49 countries a comprehensive overview of publications on the topic of the DEA, see [96].

do not have to be known a priori but are determined by the data of the inputs and outputs. Thus, for each DMU_i (i = 1, …, I) the weights can be calculated, on the one hand aiming to ensure that input and output factors are condensed into real-valued quantities ($\mathbb{R}$) and, on the other hand, maximizing their own efficiency [80, 100]). Mathematically, the presented situation is formulated as an optimization problem with the goal of maximizing the efficiency value $E_{I,\,CCR}$ of the DMU_i. The set of inputs $x_{i.k}$ of all k input types, which can take the form k = (1, …, K) is weighted by the relative weight w_k. The set of outputs $y_{i.j}$ of all j output types, which can take the form j = (1, …, J), is weighted by the relative weight v_j. The products of the input sets and their relative importance weights, yield the virtual value output of the DMU.

$$\text{virtual outputs} = v_j y_{1.j} + \cdots + v_J y_{I.K} \tag{1}$$

$$\text{virtual inputs} = w_k x_{1.k} + \cdots + w_K x_{I.K} \tag{2}$$

$$\text{DEA efficiency} = \frac{\text{virtual outputs}}{\text{virtual inputs}} = \frac{v_j}{y_{1.j}} + \cdots + v_J y_{I.K} w_k x_{1.k} + \cdots + w_K x_{I.K} \tag{3}$$

Since the DEA efficiency value represents the quotient of virtual output and virtual input, this form is referred to as the quotient form and the optimization problem. By applying CRS, the following formula is called the quotient program of a CCR model. Once the objective function (4) is assessed for the performance evaluation of a DMU, the restrictions (5) and (6) are used. Equation (5) states that the calculated efficiency value must be less than or equal to 1, which makes it possible to interpret efficiency as a relative number. Equation (6) is the non-negativity condition of the weights.

$$E_0^{CCR} = \frac{\sum_{j=1}^{J} v_j y_{0.j}}{\sum_{i=1}^{m} w_k x_{0.k}} \tag{4}$$

under the restrictions:

$$\frac{\sum_{j=1}^{J} v_j y_{0.j}}{\sum_{i=1}^{m} w_k x_{0.k}} \le 1; \quad \forall j = 1, \ldots, n \tag{5}$$

$$w_k, v_i \ge 0; \quad \forall j = 1, \ldots, J; \forall k = 1, \ldots, K \tag{6}$$

Further mathematical formulations, e.g., CCR in multiplier or envelopment form as well as the BCC model [15] with variable returns to scale (VRS), are no subject of this publication.

3.2 *Efficiency Analysis in Digitalization Scenarios*

After explaining the relevant basics of DEA, a DEA based model for measuring the efficiency of work systems of professional drivers in digitalization scenarios will be developed. This is based on a case analysis within the distribution logistics of a German food retail company. The goal is to identify the DMUs, input, and output factors.

Relevant work systems of professional drivers can be found in the transport sector of a retailing company. It is responsible for delivering food and non-food items from the regional logistics center punctually and completely to all retail stores in the respective delivery area and for redistributing recyclables, empty loading equipment, and returns to the distribution center. In order to make the underlying scenario comprehensible, the following description of the supply chain of a food retailing company is divided into financial flows and physical material flows. From a financial point of view, the distribution unit of the retail company is burdened with individual service costs by the logistics sector, which counts as a source of revenue for the logistics sector. To deliver all orders, the logistics division of the trading company uses only transport companies of different sizes. The tours are planned by the dispatchers of the retailer. The transport companies receive a payment for the physical distribution of the goods, which is based on the actual transported units. For their part, the carriers must reimburse the monetary value of all goods damaged in the delivery process. Looking more closely at the truck drivers' daily work routine, it can be divided into the following steps (Fig. 6).

Aiming to develop an efficiency measurement in digitalization scenarios, the following question has to be answered: Which work process of the professional drivers is the one with a maximum of human–machine interaction? An analysis of

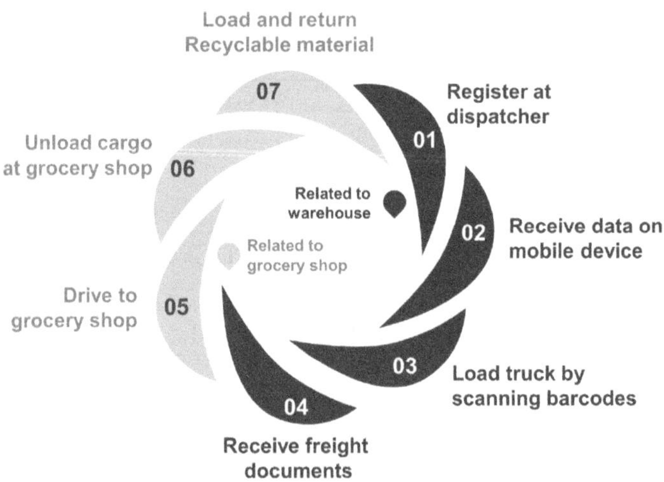

Fig. 6 The workflow of a professional truck driver in the case analysis

the entire chain is unsuitable because there are too many potential interfering factors, e.g., waiting times in the logistics center or at the retail stores, traffic jams, and vehicle failures, that can distort the time factor as an important element of efficiency analysis. The sub-processes (1), (4) and (5) are also unsuitable because they are not affected by a change in the degree of digitalization. The sub-processes (6) and (7) are suitable for the scenario but require less interaction with the mobile device than (2) and (3). The focus is, therefore, on the loading processes (2) and (3) as they take up a large part of the interaction time with the mobile device by scanning all the units to be transported. Consequently, the goal is to investigate the work system "truck loading." The human being, in this context the professional driver, affects the work object. These are the transported units loaded at the source (logistics center) and unloaded at the sink (retail shop). All loading operations must be documented with a tool, the mobile device. As input factors information, and materials flow into the work system, which produces the loading or unloading of the units to be transported as a result of work. The digital changeover within the work system is taking place through the introduction of a new handheld scanner, which covers further loading processes by digital technologies. Specifically, the Zebra MC9190-Z PDA handheld with a Microsoft Windows Mobile 6.5 operating system is replaced by the Zebra TC75 based on the Android operating system. The extension within the named work system refers to the newly introduced loading documentation of high-value items as a special virtual dialogue. The previous, paper-based process will be abolished in this context. For the users of the mobile device, optical changes result from the introduction of a new operating system. In addition, users do no longer use a keyboard when operating but rather a full-touch display, which influences haptic perception. Based on the identified work system and the defined DMUs, the loading time of the truck and the incurred transport costs are used as input for the DEA basic model. The loading time is defined as the time difference between taking over the tour by the driver and the systemic confirmation of the loading end. The second factor is quantified by the remuneration of the transport companies for the transported units. However, the use of the entire freight tariff is ineffective because only one sub-process, the loading process, is analyzed. Therefore, only a percentage of the total compensation, based on the ratio of required travel time and total tour time, is used. The total time is based on the target time calculation of the route planning software and not on the actual total time required, in order to exclude confounding factors in the form of additional times due to traffic obstructions, traffic jams or waiting time from consideration. The parameter is calculated as follows:

$$\text{Relevant transportation costs} = \frac{\text{Actual loading time}}{\text{Planned total time of tour}} \times \text{Total payment}$$

Three factors are used as outputs. The first one is the invoice of the transportation unit of the retailing company to the sales unit for all rendered logistics services calculated by applying the equal percentage as for the input transportation costs. It can be seen as revenue for the logistics sector which is often used as an output factor for performance evaluation of logistics firms based on a DEA model [47, 77, 95,

106]. The second output factor is the monetary value of all goods that are damaged or destroyed while loading the truck. In this case, the sales prices of goods that can be found in the grocery shops are not used, but the purchase price of the food retailing company which presents the product value at the time of damage. As high output values increase the efficiency of DMUs, the reciprocal calculated by the following formula is used.

$$\text{Damage value of DMU} = \text{Max. observed damage} - \text{damage caused by DMU}$$

The third output factor is related to the transport units that are loaded into the truck. Simply using the number of units and arguing that when loading a roller container (square measure 72 cm × 81,5 cm and max. load 500 kg) or a bigger pallet (square measure: 120 cm × 80 cm and max. load 1500 kg) one barcode scan is required , and therefore the effort is equal, seems too vague. With regard to the driver loading his truck, the different weights of the transport units cause a fundamental difference in the effort of handling. Therefore, this unequal effort is taken into account by rating loading equipment based on their geometrical square measure. 1.0 transport unit (TU) shall correspond to the dimensions of a roller container and all others are converted based on this standard, e.g. 1.64 TU for a pallet [(120 cm × 80 cm)/(72 cm × 81,5 cm)]. Figure 7 summarizes the components of the efficiency analysis in digitalization scenarios and answers RQ_1 "What are the components of an efficiency analysis aimed at evaluating the effect of changing levels of digitalization within the working systems of professional truck drivers?".

The following section will present, analyze, and discuss the results of two case analyses that apply the input-oriented CCR model elaborated in this chapter.

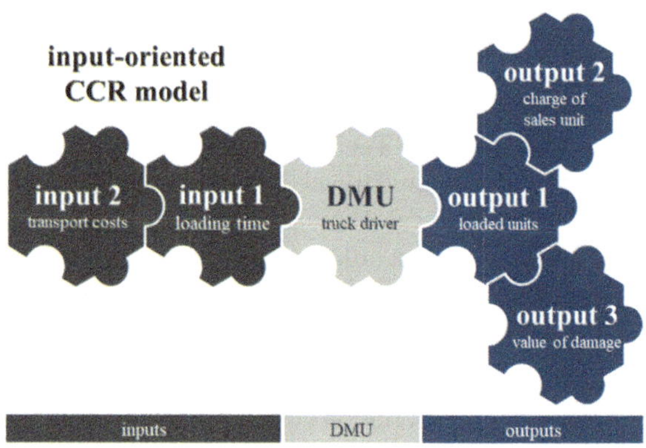

Fig. 7 Components of the efficiency analysis

4 Empirical Result Analyses

Four transport logistics centered case analyses (C1–C4) examine the efficiency of truck drivers working in the sector of distribution logistics for a large German food retailing company. Table 1 summarizes the details of the case analyses.

4.1 Efficiency Analysis for Truck Drivers in C1 and C2

The first case analysis (C1) investigated the work system of truck loading for professional truck drivers in retail logistics for 4.5 weeks and aimed to evaluate the efficiency progression of a digital changeover. The examination analyzed data of approximately 1,350 delivery tours and focused on a changing level of the digitalized work equipment within retail logistics in the described scenario. A n = 15 group experiencing a change in the level of digitalization and a n = 15 control group was examined. The model uses input and output factors described in the previous chapter and applies an input-oriented CCR model. CRS is used due to the following reasons: (1) The results of the CCR and the BBC models are similar, (2) the analysis does not particularly search for increasing or decreasing returns to scale and (3) it is assumed that the MPSS in C3, C4 and C5, the individual performance capability of professional truck drivers, is equal [15].

Table 2 illustrates the arithmetic mean values for the two examination groups as a longitudinal study, where 4.5 weeks are divided into 9 periods with 3 days each. Through the DEA analysis it was possible to gain practical knowledge about the development of efficiency when the level of digitalization increased within working systems of retail logistics. Figure 8 summarizes the structure of the study groups and the empirical curve progression of C1. Even though the empirical curve progression reflects the theoretical curve progression of Lewin (Sect. 2.2), it was not possible to prove long-term efficiency improvements.

As C1 was not able to answer RQ_3, C2 was conducted during 9.5 weeks, analyzing data of approximately 2,100 delivery tours. Table 3 summarizes relevant results.

Figure 9 summarizes the structure of the study groups and the empirical curve progression of C2. It was possible to verify the findings of C1 when reflecting the

Table 1 Sample size and duration of the case analyses

Case	n	Object of research	Duration
C1	30	Digitalization within the work system of truck loading	4.5 weeks
C2	30	Extension C3 (enlargement of duration)	9.5 weeks
C3	30	Extension C4 (enlargement of model)	9.5 weeks
C4	15	Extension C3 (enlargement of duration)	5 months

Table 2 Results of DEA basic model in C1

Period	Min	Mean	Max	Min	Mean	Max	Diff
	DG			CG			
t_{-4}	0.80	0.93	1.00	0.64	0.94	1.00	+0.01
t_{-3}	0.78	0.93	1.00	0.68	0.94	1.00	+0.01
t_{-2}	0.79	0.94	1.00	0.86	0.94	1.00	±0.00
t_{-1}	0.72	0.92	1.00	0.85	0.93	1.00	+0.01
t_0	0.63	0.80	1.00	0.61	0.86	1.00	+0.06
t_{+1}	0.74	0.89	1.00	0.76	0.91	1.00	+0.02
t_{+2}	0.77	0.93	1.00	0.76	0.92	1.00	−0.01
t_{+3}	0.67	0.89	1.00	0.66	0.91	1.00	+0.02
t_{+4}	0.78	0.92	1.00	0.69	0.88	1.00	−0.04

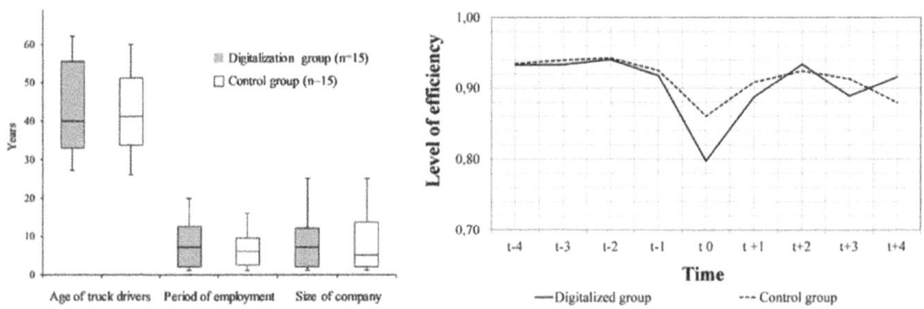

Fig. 8 Structure of the study groups and empirical curve progression of C1

theoretical curve progression presented in Sect. 2.2. and, furthermore, prove long-term efficiency improvements.

The results of C1 and C2 answer RQ_2 "Which level of efficiency exists before the change in the level of digitalization, how does it develop during and after the changeover?" as follows: Before the change in the level of digitalization, the level of efficiency is constant, during the digital changeover the level of efficiency is falling considerably and surpassing the average level after four weeks. Regarding RQ_3 "Does the efficiency analysis determine an efficiency gain or a loss of efficiency for the examined working systems?" it can be stated that C2 was able to prove a gain of efficiency, whereas C1 was not.

Table 3 Results of DEA basic model in C2

Period	Min	Mean	Max	Min	Mean	Max	Diff
	DG			CG			
t_{-8}	0.81	0.93	1.00	0.69	0.90	1.00	+0.03
t_{-7}	0.77	0.94	1.00	0.72	0.92	1.00	+0.02
t_{-6}	0.78	0.93	1.00	0.76	0.92	1.00	+0.01
t_{-5}	0.80	0.94	1.00	0.75	0.92	1.00	+0.02
t_{-4}	0.72	0.92	1.00	0.64	0.92	1.00	−0.01
t_{-3}	0.77	0.93	1.00	0.72	0.93	1.00	±0.00
t_{-2}	0.67	0.92	1.00	0.83	0.91	1.00	+0.01
t_{-1}	0.72	0.92	1.00	0.82	0.90	1.00	+0.02
t_0	0.70	0.86	1.00	0.79	0.92	1.00	−0.06
t_{+1}	0.71	0.89	1.00	0.77	0.93	1.00	−0.04
t_{+2}	0.74	0.87	1.00	0.64	0.86	1.00	+0.01
t_{+3}	0.69	0.87	1.00	0.71	0.86	1.00	+0.01
t_{+4}	0.66	0.83	1.00	0.70	0.86	1.00	+0.01
t_{+5}	0.73	0.91	1.00	0.69	0.83	1.00	+0.08
t_{+6}	0.77	0.90	1.00	0.74	0.87	1.00	+0.03
t_{+7}	0.74	0.96	1.00	0.75	0.90	1.00	+0.06
t_{+8}	0.88	0.95	1.00	0.84	0.90	1.00	+0.05
t_{+9}	0.95	0.99	1.00	0.81	0.92	1.00	+0.07
t_{+10}	0.98	0.99	1.00	0.68	0.86	1.00	+0.13

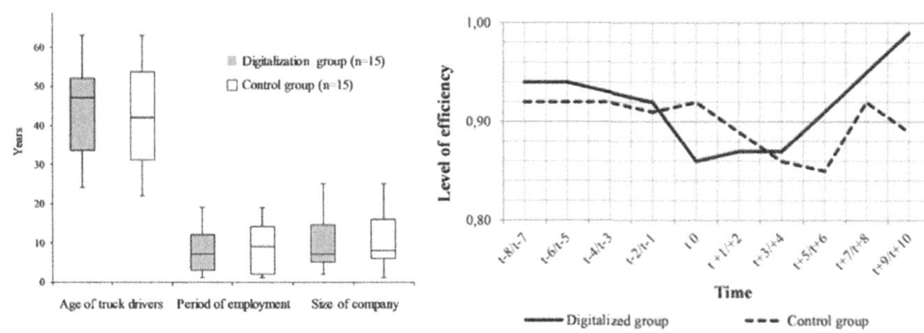

Fig. 9 Structure of the study groups and empirical curve progression of C2

4.2 Extension of the DEA Model in C3

Based on the validated CCR base model, several extensions are now examined by testing quantitative additions of input or output factors. Each enhancement is considered to be of value as long as it stabilizes the validated efficiencies of the CCR base

Table 4 Results of extensions 1–4 in C3

	Min	Mean	Max	DMU	Min	Mean	Max	DMU
	DG				CG			
B	0.74	0.96	1.00	4	0.75	0.90	1.00	3
B + (1)	0.75	0.96	1.00	5	0.75	0.90	1.00	3
B + (2)	0.75	0.96	1.00	5	0.75	0.90	1.00	3
B + (3)	0.81	0.97	1.00	5	0.77	0.91	1.00	4
B + (4)	0.75	0.96	1.00	4	0.75	0.90	1.00	3

model and does not fundamentally change them. At the same time, the results of the extensions must withstand a critical discourse with current research results and, if necessary, modifications must be worked out and discussed. The efficiency values of the period t_{+7} are used as a reference framework (basic model = B), whereby a change or stabilization of the efficiency values is examined by the behavior of the minimum, average and maximum values as well as by the number of efficient DMUs. The first approach is to extend the loaded units by the transport weights. These may be included in isolation as another output factor, shown as extension (1), or also in the form of a point system illustrated by extension (2). Furthermore, it can be argued that the 1:1 relationship between the loading gate and the transported units per customer results in additional expenses when loading more than one customer. Therefore, the number of customers is included in the extension (3) in isolation. In extension (4), a point system of the area, weight, and customer is created and applied (Table 4).

The four expansions show different effects on the minimum, average, and maximum values. Equations (1), (2), and (4) stabilize the values of the CCR base model, with (1) and (2) producing more efficient DMUs. Equation (3) is not suitable due to the resulting deviations. Only case (4) can be considered as a value-added contribution since the point system for the inclusion of loaded units, weights, and stops stabilizes the validated CCR base model and do not fundamentally change it. Since the extension of the loading effort within the working system of professional drivers represents a special concretization of commercial logistics activities, no relevant publication for a critical discourse was found.

The integration of qualitative factors has been discussed in the DEA literature since 1990, which has produced different approaches [20, 27].[9] Several publications indicate that it is not suitable to integrate qualitative within existing, solely quantitative DEA models [90, 109]. The approach of Sherman and Zhu [90] included the service quality of banks in one DEA model. Their first model calculates the efficiency of the banks without qualitative factors, while the second model quantifies the qualitative factors and integrates them as output into the calculation. Because the DMUs have different efficiency values, an isolated view of the quality occurs, which ultimately leads to the formulation of the Q-DEA model. A further refinement of this is the QE-DEA model by Zervopoulos and Palaskas [109]. The publication by

[9]For an overview see [83].

Sherman and Zhu [90] thus confirms the results of this work. Similarly, the study by Shimshak and Lenard (2007), when examining the efficiency of nursing homes, concludes that the simple integration of qualitative factors into an existing, purely quantitative DEA model leads to changes in efficiency values. Based on this, the authors developed the Two-Model Approach (TMA), in which business efficiency (X-axis) and qualitative efficiency (Y-axis) are examined separately and combined in a single portfolio. Depending on the location of the DMUs in one of the four quadrants, the overall efficiency of the examination unit is obtained. Therefore, the logic of two separate models is applied in this contribution. For the calculation of the operational efficiency, the output factors of the CCR basic model including the extension by (4) are used, whereas the qualitative efficiency is derived from the inverse value calculations of "operator error of the DMU" and "mistake in a dialog of the DMU "as output factors. In both DEA models the identical input values "transport costs" and "loading time" are used. In addition to the formulation of the two DEA models, a more desirable and identical threshold has to be set in the calculations. The portfolio matrix is primarily created for the digitalization group in order to be able to investigate the efficiency of this group in detail before, during, or after a change in the degree of digitalization. As a threshold value, this contribution determines the mean value of the economic efficiency of the CG, which is 90%. The threshold of qualitative efficiency is identical, with 90%. If a DMU is 100% efficient in only one of the two models, it must be eliminated in the following recalculation. In the second calculation cycle the remaining DMUs are based only on the 100% efficient DMUs in both models, thus avoiding a negative falsification of the result set (Table 5).

The results of the calculation, as well as the results by Shimshak and Lenard, show that the efficiency values of the two models differ. Economically efficient DMUs such as DG_{07}, DG_{10}, DG_{12}, and DG_{15} have lower qualitative efficiency. Using the TMA, high performers such as DMU DG_{05} and DG_{15} can be identified. Figure 10 visualizes the results of the TMA using a portfolio matrix.

In the portfolio chart the high-performers can be found in Q2, working economically and qualitatively efficiently. The DMU D03 in Q1 causes few error messages and false scans, but at the same time is less operational efficient than the mean value of the control group (threshold value 90%). In this case, the professional driver understands the pure interaction with the new mobile device. However, loading processes seem to be unknown and require further training. The DMUs in Q4 know the processes well, but have problems interacting with the mobile device. It is important to note that the majority of DMUs are in Q4, which may also mean that the introduced system is not intuitive enough to handle without detailed prior instruction. In addition, no DMU can be found in Q3, which would require training in procedural and systemic terms. In conclusion, it can be stated that qualitative factors can be included in the DEA calculations using the TMA of Shimshak and Lenard. It should also be noted that the inclusion of these factors is only possible for a certain period and cannot reflect the entire study period. Thus, the base model can show the economic efficiency course over n periods, and the TMA gives, based on this, in a certain period, on the existing need for retraining for professional drivers.

Table 5 Results of extensions by applying TMA in C3

DMU	I_1	I_2	O_1	O_2	eff	eff
	Time	Cost	Error	Mistake	Qual	Quant
DG_{01}	4.32	220	16	12	0.92	0.84
DG_{02}	5.87	215	18	7	0.97	0.71
DG_{03}	1.95	71	13	12	0.75	1.00
DG_{04}	4.83	174	14	10	0.97	0.80
DG_{05}	4.78	167	18	7	0.99	0.92
DG_{06}	3.77	196	15	11	0.99	0.89
DG_{07}	5.58	216	17	7	1.00	0.67
DG_{08}	5.67	225	11	11	0.94	0.68
DG_{09}	7.80	350	17	3	0.99	0.42
DG_{10}	7.23	336	18	4	1.00	0.48
DG_{11}	6.00	227	18	8	0.97	0.67
DG_{12}	6.08	219	18	3	1.00	0.70
DG_{13}	6.43	305	18	2	0.95	0.54
DG_{14}	4.47	236	18	9	0.93	0.78
DG_{15}	4.42	165	18	11	1.00	0.93

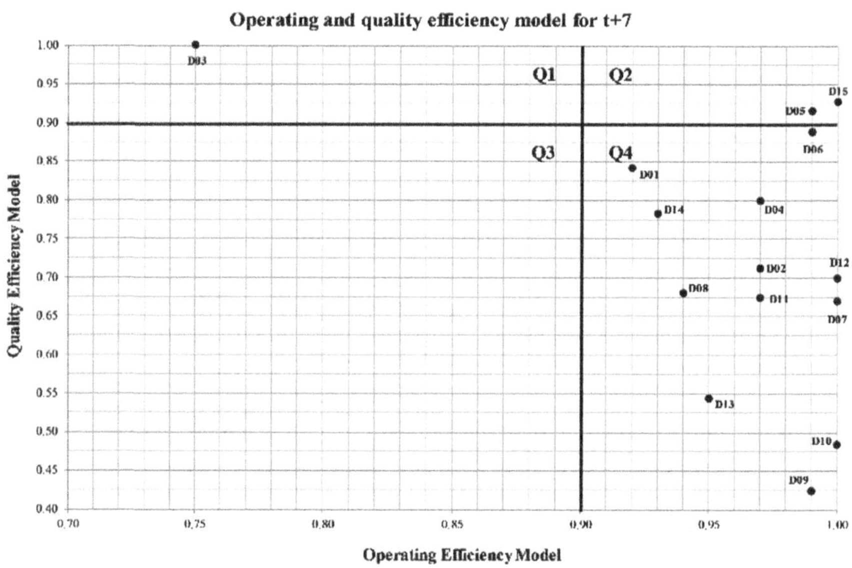

Fig. 10 Two-model approach merging qualitative and operational efficiency

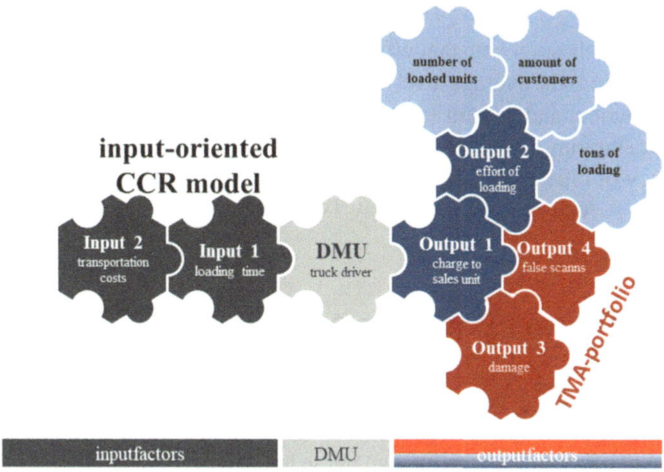

Fig. 11 Final DEA approach merging qualitative and operational efficiency

After discussing quantitative and qualitative extensions of the CCR basic model created in C1 and C2, the Fig. 11 summarizes the final DEA model created in C3, which is able to merge qualitative and operational efficiency for an efficiency analysis in digitalization scenarios.

4.3 *Enlargement of Duration in C4*

The last case analysis C6 is designed as a longitudinal study applying the DEA model that has been developed in C1 and C2 with a dataset containing approximately 112,000 transport units and 3,000 delivery tours for five months. Aiming to examine the usability of long-term efficiency measurement, the DG was measured after period t_{+10} in C2. Figure 12 summarizes the efficiency progression of the n = 15 truck drivers

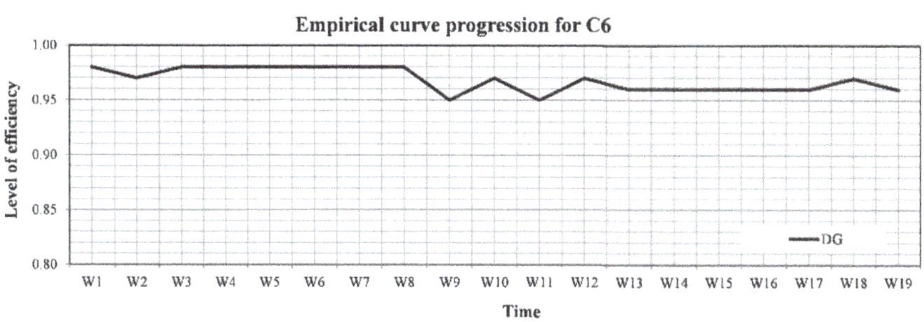

Fig. 12 Empirical curve progression for DG during 10 weeks in case analysis C4

of DG.

The analysis shows a constant efficiency level between 0.98 (e.g., t_{W1}) and 0.95 (e.g., t_{W9}), whereas the average level of efficiency on a lower after t_{W9}. As the analysis starts in September and ends in January, two weeks before Christmas were excluded due to high shipping volume. The seasonal effects in full-range food retailing on the efficiency of blue-collar is, furthermore, the reason for the volatile development from t_{w9} to t_{w11} (German public holiday 1st November).

5 Correlation and Regression Analysis

Aspiring to answer RQ_4 "What effect do demographic factors such as age, employer size, period of employment, migration background and job profile of parents have on the level of efficiency after a change in the level of digitalization?", several correlations and regression analyses are conducted. The correlations coefficient r for the relationship of "age in years" (x) and "level efficiency" (y) is $r = 0.1496$. Therefore, there is a low statistical linear relationship between these 2 factors. Figure 13 shows the correlation graphs for the relationship of "total trucks of the company" for the carrier's size (x) and "level of efficiency" (y) as well as the relationship of "period of employment" at the food retailing company and not as a truck driver (x) and "level of efficiency" (y).

Due to the correlation coefficients, there is a strong positive linear statistical relationship between a company's size and the level of efficiency during a digital change. Furthermore, a strong positive linear statistical relationship between the period of employment in a certain occupation and the level of efficiency during a digital change can be determined. Figure 14 summarizes the results and methods for

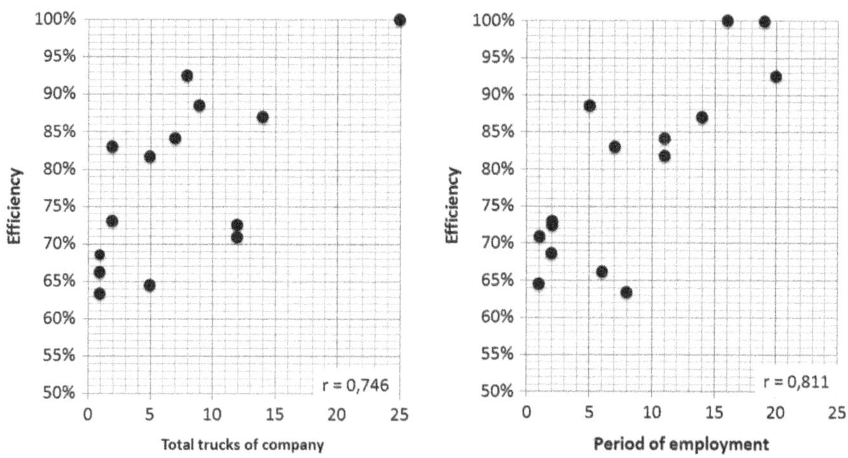

Fig. 13 Results of correlation analyses for company size and period of employment

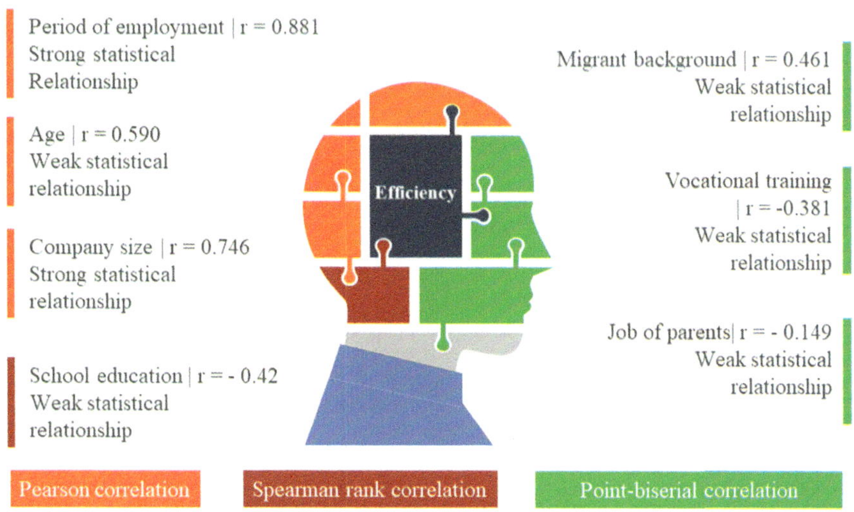

Fig. 14 Results of correlation analyses for all demographic factors

all correlation analyses conducted.

As correlation analyses do not distinguish between dependent and independent variables, nine regression analyses are conducted in order to examine the relationship to exogenous factors integrated into the work system, which are determined as characteristics of the employees: age (interval scale), seniority (interval scale), company size (ordinal scale), migration (dichotomous y/n), education (ordinal scale), vocational training (ordinal scale) and job of parents (dichotomous, truck driver y/n). Further endogenous factors that are resulting out of the digital changeover, such as satisfaction and motivation are analyzed (both as interval scale from the questionnaire). Table 6 summarizes the results.

The results show that the seniority of truck drivers has a high influence on the level of efficiency during a digital changeover ($R^2 = 0.864$), whereas the other exogenous factors have no influence in this scenario. Regarding the endogenous factors, it can be stated that the truck drivers' individual perception of motivation and satisfaction does not affect the truck drivers' performance.

6 Implications for Management

The following chapter uses (1) the empirical curve for the level of efficiency after the digital changeover examined in C1–C2 (Sect. 4) as well as (2) the correlation and regression analyses (Sect. 5) in order to derive relevant implications for management regarding logistics companies, personnel management, and software development. During the whole elaboration, great importance is attached to the fact that high

Table 6 Results of regression analysis for exogenous and endogenous factors on efficiency

	Dependent variable: efficiency								
Age	−0.0004								
	(0.003)								
Seniority		0.015^{***}							
		(0.002)							
Company size			−0.022						
			(0.031)						
Migration				−0.011					
				(0.088)					
Education					−0.028				
					(0.048)				
Vocational training						0.139^{*}			
						(0.064)			
Job of parents							0.071		
							(0.058)		
Satisfaction								−0.010	
								(0.006)	
Motivation									0.014
									(0.014)
Constant	0.876^{***}	0.708^{***}	0.911^{***}	0.866^{***}	0.899^{***}	0.829^{***}	0.828^{***}	1.091^{***}	0.563^{*}
	(0.113)	(0.020)	(0.081)	(0.082)	(0.079)	(0.029)	(0.037)	(0.141)	(0.286)
Observ	15	15	15	15	15	15	15	15	15

(continued)

Table 6 (continued)

	Dependent variable: efficiency								
R^2	0.002	0.864	0.038	0.001	0.025	0.263	0.104	0.182	0.076
Adjusted R^2	-0.074	0.854	−0.036	−0.076	−0.050	0.206	0.035	0.119	0.005
Residual std. error	0.116	0.043	0.114	0.116	0.115	0.100	0.110	0.105	0.112
F statistic	0.030	18.693***	0.514	0.016	0.338	4.643*	1.504	2.891	1.069

Note *p **p ***p < 0.01

correlation values stating a high linear statistical relationship are not equated with causality. This is crucial when it comes to an assessment of the explanatory power of this contribution.

6.1 Implications for Logistics Companies

For (1), no cause-and-effect relationships can be derived, because there is the danger that too many disturbance variables are present within a too small number of cases. The causal relationship "if a company increases its degree of digitalization (cause), it leads to a loss of efficiency (effect), after the changeover" does not withstand any critical discourse with already published publications on this topic. Helfrich solves this kind of problem by stating that target-means relationships are derivable even without causal statements when validating or evaluating the results [44]. If a nomological statement is replaced by a nomopragmatic statement in the sense of a more concrete statement, action-relevant instructions can be derived, which can withstand a critical discourse. Therefore, the findings of this work allow the following nomopragmatic statement: "If within the power range of distribution logistics, the level of digitalization within work systems of professional drivers is increased and they are briefed only in the functioning of the mobile device (cause), then this leads to a loss of efficiency (effect) directly after the digital changeover". For the range of services provided by logistics companies, the following recommendations for action can be formulated: Either a temporary loss of efficiency in the type of change management processes must be taken into account, or suitable further training measures for professional drivers must be developed and tested. For (2), the correlation analysis, which states a statistical correlation between efficiency level and length of service, merely indicates that a short period of employment can lead to a low level of efficiency in digital change processes. Thus, there is the possibility that logistical areas of high employee turnover among professional drivers are more affected by efficiency losses during the changeover phase than those that can be attributed to long-term employees.

6.2 Implications for Human Resource Management

In (1), the professional driver, as a person within a working system, is of particular importance. In order to derive relevant implications for personnel management, an approach from sociological research that illustrates the effects of digital transformation within a socio-technical system is applied. See and Kersten differentiate between the dimension of leadership (vision and strategic guidelines, leadership style) and qualification (competence profiles, further education) [89]. The organizational dimension is divided into the management of value chains (roles and tasks, transparency) as well as an internal organization (working methods and culture, structures of cooperation). The loss of efficiency after the change process is relevant for the qualification dimension as it has to develop appropriate training concepts for the further education of professional drivers, which either prepares the employees for a digital transformation or accompanies them during this time. The competence profiles of the "blue-collar workers" have not changed in the present change process. In contrast, the observed effects on the second group of employees, only indirectly affected by the change process, are interesting. The increasing technological complexity of the professional drivers' work system affects white-collar workers, more precisely, qualified warehouse personnel, and dispatchers. Due to the increasing digitalization and interaction with high-tech systems, they face new challenges. In the case analyses, it was observed that system crashes of the mobile device could easily be handled by qualified warehouse personnel and dispatchers in the old system. With the increase in digitalization, less than half of all cases were resolved on-site, causing increasing commissioning of the IT department of the retailing company. With a steadily increasing degree of digitalization, future skill requirements for qualified warehouse personnel and dispatchers could be IT skills that go beyond mere user knowledge. Further competence requirements of qualified specialist work in logistics could also be the targeted evaluation and interpretation of data, which are becoming increasingly available as a result of the digital transformation. The impact on the organizational dimension has not been addressed in this contribution. For (2), the result of the correlation and regression analyses show a linear statistical relationship between the two characteristics efficiency level and length of service, which is a reason for further derivation. Although no causal link can be stated, the results indicate that the process safety of the professional drivers within their work system may be of key importance to the success of a digital change process. Employment of more than five years and the knowledge gained during this time seem to minimize the loss of efficiency during a digital change process. This means that haulage managers' personnel management should focus on developing and promoting employee retention programs to reduce the turnover of professional drivers. Clients, in this case the transport sector of the food retailing company, can contribute to this with a positive working environment, e.g., short waiting times, clear objectives, and permanent support of professional drivers.

6.3 Implications for Software Development

Regarding (1), software companies can use the TMA's portfolio to measure and, if necessary, redefine the degree of success of user-device interaction. If a large proportion of the DMUs are in Q4, they will be economically efficient with low quality-centered efficiency values at the same time. Therefore, it can be derived that the software is not intuitive enough for the user group, or selection fields cannot be sufficiently separated from each other. On the other hand, if there are only a few DMUs in Q4, software guides for knowledge transfer or training concepts can be checked for suitability. For (2), the results of the correlation analysis show that the age of professional drivers and their origin play only a minor role in the study of altered levels of digitalization. One reason for this is that mobile devices, e.g., mobile phones or tablets, regardless of age, background, and social status have already become part of our everyday lives. About ten years ago, the use of the internet and mobile devices was interpreted as a kind of border that had to be consciously crossed by the user. This way of thinking can be understood by the motto of the internet conference re:publica in 2007: "Life in the Net". Today, neither digital end devices nor the internet are parallel worlds in which users have to consciously intervene. Rather, they have become a medium that can permanently and often unconsciously expand the reality in which we live. All professional drivers examined in this contribution were in possession of a private smartphone and familiar with the operation of a full-touch display. One recommendation for a software company is to base virtual dialogs on mobile devices on the visualization and operating logic of popular smartphone models.

7 Conclusion and Outlook

The efficiency analysis is able to measure changes in the degree of digitalization within the professional drivers' work systems from an a posteriori perspective. The method developed in this contribution is characterized by the fact that efficiency progressions can be measured over a certain period of time (empirical curve) and, at the same time, the economic, as well as qualitative efficiency level of individual DMUs can be illustrated (TMA portfolio). The examination shows that the digitalization of professional truck drivers' work systems can lead to greater efficiency. Possible starting points for further methodological developments are based on the examination of other areas of activity, groups of employees, model variants, and experimental setups. It is conceivable to observe professional drivers who work in commercial vehicle fleets and not in transport companies, as well as to conduct a comparison of drivers from several depots or a comparison of depots in different regions (national and international). The use of the model should continue to be examined for work systems of professional drivers outside the distribution logistics of trading companies, e.g., for industrial works or postal and courier services. Furthermore, within the group of "blue-collar workers" activities, such as picking and in-house transport, changes in the efficiency of digitalization projects have to be investigated. Furthermore, in comparison with the group of white-collar workers, similarities, and/or differences between the two groups of employees can be derived. On the other hand, research contributions should also place further additions to the input and output factors with additional quantitative, qualitative, and soft components at the center of future considerations as well as examine the potential of the DEA Window Analysis in this context. Finally, further experimental setups could investigate the impact of training in the form of, e.g., seminars, self-studies, or digital mock-up before increasing the degree of digitization. Figure 15 summarizes further research needs clustered in four fields.

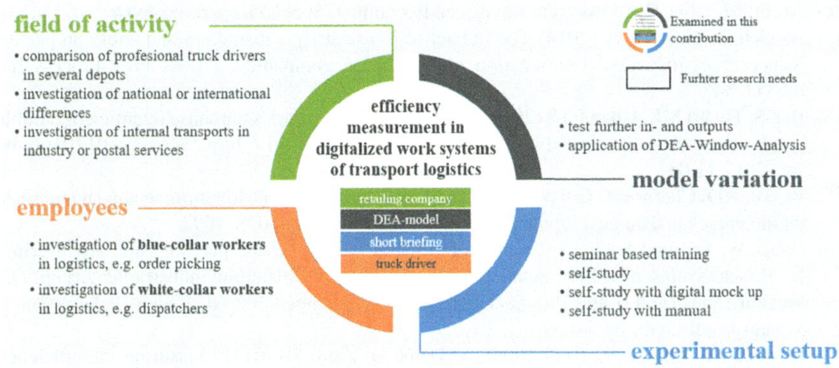

Fig. 15 Further research needs

Overall, the efficiency analysis and the management implications presented made clear that digitalization cannot be reduced to a technical aspect. It is a form of transformation that permeates the concept of work systems and can be measured using a quantitative-empirical approach. For the future of logistics, evaluation models of this kind will be important in academic and practical questions.

References

1. Ahrens D (2016) Neue Anforderungen im Zuge der Automatisierung von Produktionsprozessen: Expertenwissen und operative Zuverlässigkeit. Arbeits und Industriesoziologische Studien 9:43–56
2. Aigner D, Lovell K, Schmidt P (1977) Formulation and estimation of stochastic frontier production function models. J Econ 6:21–37. https://doi.org/10.1016/0304-4076(77)90052-5
3. Al-Fuqaha A, Guizani M, Mohammadi M, Aledhari M, Ayyash M (2015) Internet of things: a survey on enabling technologies, protocols and applications. IEEE Commun Surv Tutor 17:2347–2376. https://doi.org/10.1109/COMST.2015.2444095
4. Alter S (1995) How should business professionals analyze information systems for themselves. In: Working conference on basic information system concepts, pp 284–299
5. Alter S (1999) A general, yet useful theory of information systems. Commun Assoc Inf Syst 1:1–70
6. Alter S (2000) Same words, different meanings: are basic IS/IT concepts our self-imposed tower of babel. Commun Assoc Inf Syst 3:1–87
7. Alter S (2008a) Defining information systems as work systems: implications for the IS field. Euro J Inf Syst 17:448–469
8. Alter S (2008b) Service system fundamentals: work system, value chain, and life cycle. IBM Syst J 47:71–85
9. Alter S (2009) Bridging the chasm between socio-technical and technical views of systems in organizations. In: Working papers on information systems, vol 9
10. Alter S (2013) Work system theory: overview of core concepts, extensions, and challenges for the future. J Assoc Inf Syst 14:72–121
11. Atzori L, Iera A, Morabito G (2010) The internet of things: a survey. Comput Netw 54:2787–2805. https://doi.org/10.1016/j.comnet.2010.05.010

12. Avent R (2014) The third great wave. The Economist, Special Report, pp 1–14
13. Azadeh A, Mousavi A (2014) The impact of job security, satisfaction and stress on performance assessment and optimization of generation companies. J Loss Prev Process Ind 32:343–348
14. Bag S, Tiwari MK, Chan FTS (2019) Predicting the consumer's purchase intention of durable goods: an attribute-level analysis. J Business Res 94:408–419. https://doi.org/10.1016/j.jbusres.2017.11.031
15. Banker RD, Charnes A, Cooper WW (1984) Some models for estimating technical and scale inefficiencies in data envelopment analysis. Manage Sci 30:1078–1092
16. Bauer W, Schlund S (2015) Change of work in indirect areas: planning and engineering. In: Hirsch-Kreinsen H, Ittermann P, Niehaus J (eds) Digitalisierung industrieller Arbeit: Die Version Industrie 4.0 und ihre sozialen Herausforderungen, 1st ed. Baden-Baden; Nomos Verlagsgesellschaft, pp 54–69
17. Bayraktar E, Tatoglu E, Turkyilmaz A, Delen D, Zaim S (2012) Measuring the efficiency of customer satisfaction and loyalty for mobile phone brands with DEA. Expert Syst Appl 39:99–106. https://doi.org/10.1016/j.eswa.2011.06.041
18. Boes A, Kämpf T, Lühr T, Marrs K (2014) Kopfarbeit in der modernen Arbeitswelt: Auf dem Weg zu einer Industrialisierung neuen Typs. In: Sydow J, Sadowski D, Conrad P (eds) Arbeit—eine Neubestimmung: Managementforschung 24. Gabler, Wiesbaden, pp 33–42
19. Bogner DP (2017) Die Feldtheorie Kurt Lewins: Eine vergessene Metatheorie für die Erziehungswissenschaft. Springer Fachmedien, Wiesbaden
20. Bowen WM (1990) Subjective judgements and data envelopment analysis in site selection. Comput Environ Urban Syst 14:133–144. https://doi.org/10.1016/0198-9715(90)90018-O
21. Broy M, Geisberger E (2012) AgendaCPS: Integrierte Forschungsagenda Cyber-Physical Systems. Acatech-Studie. Springer, Berlin, Heidelberg
22. Brynjolfsson E, McAfee A (2014) The second machine age: work, progress, and prosperity in a time of brilliant technologies. W.W. Norton & Company, New York
23. Charnes A, Cooper WW, Rhodes E (1978) Measuring the efficiency of decision making units. Eur J Oper Res 2:429–444. https://doi.org/10.1016/0377-2217(78)90138-8
24. Charniak E, McDermott D (1985) Introduction to artificial intelligence. World student series edition. Reading: Addison-Wesley
25. Chen Z, Matousek R, Wanke P (2018) Chinese bank efficiency during the global financial crisis: a combined approach using satisficing DEA and support vector machines. North Am J Econ Fin 43:71–86
26. Chen Z, Wanke P, Antunes JJM, Zhang N (2017) Chinese airline efficiency under CO_2 emissions and flight delays: a stochastic network DEA model. Energy Econ 68:89–108. https://doi.org/10.1016/j.eneco.2017.09.015
27. Chilingerian JA, Sherman DH (1990) Managing physician efficiency and effectiveness in providing hospital services. Health Serv Manage Res 3:3–15. https://doi.org/10.1177/095148489000300101
28. Cooper WW, Seiford LM, Zhu J (eds) (2011) Handbook on data envelopment analysis. Springer, Boston
29. Cooper WW, Seiford LM, Tone K (2007) Data envelopment analysis: a comprehensive text with models, applications, references and DEA-solver software. Springer, Boston
30. Debreu G (1951) The coefficient of resource utilization. Econometrica 19:273–292. https://doi.org/10.2307/1906814
31. DIN EN ISO 9000 (2000) Qualitätsmanagementsysteme: Grundlagen und Begriffe. Beuth, Berlin
32. Dombrowski U, Riechel C, Evers M (2014) Industry 4.0: the role of humans in the fourth industrial revolution. In: Kersten W (ed) Industrie 4.0: Wie intelligente Vernetzung und kognitive Systeme unsere Arbeit verändern. Gito, Berlin, pp 129–153
33. Du J-Y (2013) Staff performance appraisal based on data envelopment analysis. J Chem Pharm Res 5:102–105. Retrieved from https://www.scopus.com/inward/record.uri?eid=2-s2.0-84890951264&partnerID=40&md5=7f54fe02057fd69c177250ac8a7b7d17

34. Dugelova M, Strenitzerova M (2015) The using of data envelopment analysis in human resource controlling. Proc Econ Fin 26:468–475. https://doi.org/10.1016/S2212-5671(15)00875-8
35. Eisenmann M, Ittermann P (2017) Hybrid services and change of work: challenges and perspectives in logistics. Soziologisches Arbeitspapier 50:1–46
36. Farrell MJ (1957) The measurement of productive efficiency. J R Stat Soc 120:253–290
37. Forschungsunion (2012a) Im Fokus: Das Zukunftsprojekt Industrie 4.0: Handlungsempfehlungen zur Umsetzung. Berlin
38. Forschungsunion (2012b) Umsetzungsempfehlungen für das Zukunftsprojekt Industrie 4.0: Abschlussbericht des Arbeitskreises Industrie 4.0. Berlin
39. Forschungsunion and Acatech (2013) Deutschlands Zukunft als Produktionsstandort sichern: Umsetzungsempfehlungen für das Zukunftsprojekt Industrie 4.0. Frankfurt: Abschlussbericht des Arbeitskreises Industrie 4.0
40. Glistau E, Machado NIC (2018) Industry 4.0, logistics 4.0 and materials—chances and solutions. Mater Sci Forum 919:307–314. https://doi.org/10.4028/www.scientific.net/MSF.919.307
41. Gubbi J, Buyya R, Marusic S, Palaniswami M (2013) Internet of things: a vision, architectural elements and future directions. Future Gener Comput Syst 29:1645–1660. https://doi.org/10.1016/j.future.2013.01.010
42. Hardenacke H, Peetz W, Wichardt G (1985) Work science. Studienbücher der Wirtschaft. Hanser, München
43. Haugeland J (1985) Artificial intelligence: the very idea, vol 151. MIT Press, Cambridge
44. Helfrich H (2015) Wissenschaftstheorie im betriebswirtschaftlichen Studium. Springer, Wiesbaden
45. Hirsch-Kreinsen H (2015) Gestaltungsperspektiven von Produktionsarbeit bei Industrie 4.0. In: Schlick C (ed) Arbeit in der digitalisierten Welt: Beiträge der Fachtagung des BMBF 2015, 1st ed. Frankfurt am Main: Campus
46. Hirsch-Kreinsen H, ten Hompel M, Ittermann P, Dregger J, Niehaus J, Kirks T, Mättig B (2018) Social manufacturing and logistics: Arbeit in der digitalisierten Produktion. In: Wischmann S, Hartmann E (eds) Zukunft der Arbeit: Eine praxisnahe Betrachtung. Springer Vieweg, Berlin
47. Hong S, Ahmad A (2016) An explanatory review of mandatory task. In: 2nd international symposium on agent, multi-agent systems and robotics, pp 51–56. Retrieved from https://ieeexplore.ieee.org/abstract/document/7810002/
48. Huang T-H, Chen K-C, Lin C-I (2018) An extension from network DEA to copula-based network SFA: evidence from the U.S. commercial banks in 2009. Quart Rev Econ Fin 67:51–62
49. Huchler N (2016) The 'role of humans' in Industry 4.0: technically centered versus human-centered approach. Arbeits-Und Industriesoziologische Studien 9:57–79
50. Jacintho JC, Da Silva MT, Do Nascimento RJ, Poveda PF, Cevoli TN, Ribeiro VH (2018) Ethanol loading and dispatch operation: a discussion on management practices and logistics 4.0. In: ILS 2018—information systems, logistics and supply chain, proceedings. Retrieved from https://www.scopus.com/inward/record.uri?eid=2-s2.0-85050973246&partnerID=40&md5=8904e770e1c7a38c26b1f834e281f8e4
51. Kagermann H, Lukas W-D, Wahlster W (2011) Industrie 4.0: Mit dem Internet der Dinge auf dem Weg zur 4. industriellen revolution. VDI Nachrichten, 13
52. Kagermann H, Wahlster W, Helbig J (2013) Umsetzungsempfehlungen für das Zukunftsprojekt Industrie 4.0: Abschlussbericht des Arbeitskreises Industrie 4.0. Frankfurt.
53. Kerpen P (2016) Praxisorientierte data envelopment analysis. Gabler, Wiesbaden
54. Khodamoradi M, Behzadi M-H, Rostamy-MaleKhalifeh M (2016) Evaluating of staff's performance with job satisfaction and organizational commitment approach by DEA: case study: Khuzestan oxin steel company. Int J Data Envel Anal 4:945–949
55. Klumpp M (2017) Analyse und Modellierung von Wertschöpfungsfunktionen für Hochschulen. In: Zelewski S, Klumpp M, Akca N (eds) Hochschuleffizienz: Konzeptionelle Herausforderungen und Lösungsansätze aus Sicht der betriebswirtschaftlichen Forschung. Logos Berlin, Berlin, pp 125–138

56. Klumpp M (2018) Automation and artificial intelligence in business logistics systems: human reactions and collaboration requirements. Int J Logist Res Appl 21:224–242. https://doi.org/10.1080/13675567.2017.1384451
57. Klumpp M, Ruiner C (2018) Digitalization and work organization in new urban food delivery systems, pp 301–312. https://doi.org/10.18461/pfsd.2018.1823
58. Koch V, Geissbauer R, Kuge S, Schrauf S (2014) Chancen und Herausforderungen der vierten industriellen revolution. Pwc und Strategy, Frankfurt, München
59. Koopmans TC (1951) An analysis of production as an efficient combination of activities. In: Koopmans TC (ed) Activity analysis of production and allocation: proceeding of a conference. Wiley, London, pp 33–97
60. Kotter JP (1995) Leading change: why transformation efforts fail. Harvard Business Rev 73:59–67
61. Kurzweil R (1990) The age of intelligent machines. MIT Press, Cambridge
62. Lee B, Worthington AC (2016) A network DEA quantity and quality-orientated production model: an application to Australian university research services. Omega 60:26–33. https://doi.org/10.1016/j.omega.2015.05.014
63. Lee CKH (2017) A GA-based optimisation model for big data analytics supporting anticipatory shipping in Retail 4.0. Int J Prod Res 55:593–605
64. Lee E (2008) Cyber physical systems: design challenges. In: 11th IEEE symposium on object/component/service-oriented real-time distributed computing. Advance online publication. https://doi.org/10.1109/ISORC.2008.25
65. Lee J, Bagheri B, Kao H-A (2015) A cyber-physical systems architecture for industry 4.0-based manufacturing systems. Manuf Lett 3:18–23. https://doi.org/10.1016/j.mfglet.2014.12.001
66. Lewin K (1947) Frontiers in group dynamics. Human Relat 1:5–41. https://doi.org/10.1177/001872674700100103
67. Lewin K (1963) Field theory in the social sciences: selected theoretical publications. Huber, Bern
68. Loureiro RJP, Simões JMM, Cartaxo JV, Amorim MPC, Costa C, Au-Yong-Oliveira M, Amorim MPC (2018) Costs of transaction in logistics 4.0 and influence of innovation networks. In: Proceedings of the European conference on innovation and entrepreneurship, ECIE, 2018-September. Retrieved from https://www.scopus.com/inward/record.uri?eid=2-s2.0-85055425695&partnerID=40&md5=6a62298267d278d484f0fe1dd4edb671
69. Luczak H (1997) Core definition and systematics of work science. In: Luczak H, Volpert W, Müller T (eds) Handbook work science. Schäffer-Poeschel, Stuttgart, pp 11–19
70. Malhotra R, Malhotra D (2007) Using data envelopment analysis to analyze European union nations. J Int Business 2:24–39
71. Malhotra R, Malhotra D (2009) Evaluating the efficiency of European Union integration. Int J Commerce Manage 19:233–252. https://doi.org/10.1108/10569210910988001
72. Markillie P (2012) A third industrial revolution. The Economist, Special Report, pp 1–10
73. Martínez A, Schmuck C, Pereverzyev S, Pirker C, Haltmeier M (2018) A machine learning framework for customer purchase prediction in the non-contractual setting. Euro J Oper Res. Advance online publication. https://doi.org/10.1016/j.ejor.2018.04.034
74. McCarthy J, Minsky ML, Rochester N, Shannon CE (2006) A Proposal for the dartmouth summer research project on artificial intelligence. AI Magazine 27:12–14
75. Miorandi D, Sicari S, de Pellegrini F, Chlamtac I (2012) Internet of things: vision, applications and research challenges. Ad Hoc Netw 10:1497–1516. https://doi.org/10.1016/j.adhoc.2012.02.016
76. Mo Y, Kim TH-J, Brancik K, Dickinson D, Lee H, Perrig A, Sinopoli B (2012) Cyber-physical security of a smart grid infrastructure. Proc IEEE 100:195–209. https://doi.org/10.1109/JPROC.2011.2161428
77. Momeni E, Azadi M, Saen RF (2015) Measuring the efficiency of third party reverse logistics provider in supply chain by multi objective additive network DEA model. Int J Ship Transp Logist 7:21–41. https://doi.org/10.1504/IJSTL.2015.065898

78. Oleśków-Szłapka J, Stachowiak A, Burduk A, Chlebus E, Nowakowski T, Tubis A (2019) The framework of logistics 4.0 maturity model. Adv Intel Syst Comput 835:771–781. https://doi.org/10.1007/978-3-319-97490-3_73
79. Omrani H, Soltanzadeh E (2016) Dynamic DEA models with network structure: an application for Iranian airlines. J Air Transp Manage 57:52–61. https://doi.org/10.1016/j.jairtraman.2016.07.014
80. Peters ML (2017) Grundlagen der data envelopment analysis. In: Zelewski S, Klumpp M, Akca N (eds) Hochschuleffizienz: Konzeptionelle Herausforderungen und Lösungsansätze aus Sicht der betriebswirtschaftlichen Forschung. Logos Berlin, Berlin, pp 37–123
81. Poole DL, Mackworth AK, Goebel RG (1998) Computational intelligence: a logical approach. Oxford Univ. Press, New York
82. Rajkumar R, Lee I, Sha L, Stankovic J (2010) Cyber-physical systems: the next computing revolution. In: 1st design automation conference. Advance online publication. https://doi.org/10.1145/1837274.1837461
83. Richter M (2017) Time and quality efficiency of Scotch single malt whisky production: an empirical two-model DEA approach. Ilmenauer Schriften Zur Betriebswirtschaftslehre 2:1–21
84. Rifkin J (2011) The third industrial revolution: how lateral power is transforming energy, the economy and the world. Palgrave Macmillan, New York
85. Russell SJ, Norvig P (2010) Artificial intelligence: a modern approach, 3rd edn. Prentice Hall, New Jersey
86. Scheel H (2000) Effizienzmaße der data envelopment analysis. Deutscher Universitätsverlag, Wiesbaden
87. Schmidtke N, Behrendt F, Thater L, Meixner S (2018) Technical potentials and challenges within internal logistics 4.0. In: Proceedings—GOL 2018: 4th IEEE international conference on logistics operations management. Advance online publication. https://doi.org/10.1109/GOL.2018.8378072
88. Schultetus W (2006) Work science from theory to practice: work science findings and their economic benefits. Wirtschaftsverlag, Köln
89. See B von, Kersten W (2018) Arbeiten im Zeitalter des Internets der Dinge. Industrie 4.0 Manage 34:8–12. https://doi.org/10.30844/IM18-3_8-12
90. Sherman DH, Zhu J (2006) Benchmarking with quality-adjusted DEA (Q-DEA) to seek lower-cost high-quality service: evidence from a U.S. bank application. Ann Oper Res 145:301–319. https://doi.org/10.1007/s10479-006-0037-4
91. Shirouyehzad H, Hosseinzadeh LF, Aryanezhad MB, Dabestani R (2012) A Data envelopment analysis approach for measuring efficiency of employees: a case study. South African J Industr Eng 23:191–201. https://doi.org/10.7166/23-1-230
92. Stepan A, Fischer EO (2009) Betriebswirtschaftliche Optimierung (8th ed). Lehr-und Handbücher zur entscheidungsorientierten Betriebswirtschaft. Wissenschaftsverlag, Oldenbourg
93. Stowasser S, Jeske T (2015) Arbeitswelt der Zukunft. In: Schlick C (ed) Arbeit in der digitalisierten Welt: Beiträge der Fachtagung des BMBF 2015, 1st ed. Frankfurt am Main: Campus
94. Strandhagen JO, Vallandingham LR, Fragapane G, Strandhagen JW, Stangeland ABH, Sharma N (2017) Logistics 4.0 and emerging sustainable business models. Adv Manuf 5:359–369. https://doi.org/10.1007/s40436-017-0198-1
95. Tang L, Huang X, Peng Y, Xiao Z (2015) Analysis and evaluation of relative efficiency of warehousing and distribution operations based on mixed DEA model. Chem Eng Trans 46:583–588
96. Tavares G (2002) A bibliography of data envelopment analysis: Rutcor research report 1978–2001. Piscataway, Rutgers Center for Operations Research
97. Vahs D, Weiand A (2010) Workbook change-management: methods and techniques. Schäffer-Poeschel, Stuttgart
98. Van Eck NJ, Waltman L (2010) Software survey: VOSviewer, a computer program for bibliometric mapping. Scientometrics 84:523–538

99. Wee D, Kelly R, Cattel J, Breunig M (2015) Industry 4.0: How to navigate digitization of the manufacturing sector. München, Stamford, Tokyo, McKinsey & Company, Hamburg
100. Wilken R (2007) Dynamisches benchmarking: ein verfahren auf basis der data envelopment analysis. Deutscher Universitätsverlag, Wiesbaden
101. Will-Zocholl M (2016) Die Verlockung des Virtuellen: Reorganisation von Arbeit unter Bedingungen der Informatisierung, Digitalisierung und Virtualisierung. Arbeits-Und Industriesoziologische Studien 9:25–42
102. Windelband L, Dworschak B (2015) Arbeit und Kompetenzen in der Industrie 4.0: Anwendungsszenarien Instandhaltung und Leichtbaurobotik. In: Hirsch-Kreinsen H, Ittermann P, Niehaus J (eds) Digitalisierung industrieller Arbeit: Die Version Industrie 4.0 und ihre sozialen Herausforderungen, 1st ed. Baden-Baden: Nomos Verlagsgesellschaft, pp 71–86
103. Wrobel-Lachowska M, Wisniewski Z, Polak-Sopinska A, Andre T (2018) The role of the lifelong learning in logistics 4.0. Adv Intel Syst Comput 596:402–409. https://doi.org/10.1007/978-3-319-60018-5_39
104. Wróbel-Lachowska M, Wisniewski Z, Polak-Sopińska A (2018) The role of the lifelong learning in logistics 4.0. In: international conference on applied human factors and ergonomics, pp 402–409
105. Yang G-L, Fukuyama H, Song Y-Y (2018) Measuring the inefficiency of Chinese research universities based on a two-stage network DEA model. J Informetr 12:10–30. https://doi.org/10.1016/j.joi.2017.11.002
106. Ye YY (2015) Performance evaluation of logistics enterprise based on DEA model. Int J Serv Technol Manage 21:137–141. Retrieved from https://www.scopus.com/inward/record.uri?eid=2-s2.0-84957359142&partnerID=40&md5=f7db5d3faecde37cca6e57586b99527c
107. Zanella A, Bui N, Castellani A, Vangelista L, Zorzi M (2014) Internet of things for smart cities. IEEE IoT J 1:22–32. https://doi.org/10.1109/JIOT.2014.2306328
108. Zbranek P (2013) Data envelopment analysis as a tool for evaluation of employees' performance. Acta Oeconomica Et Informatica 16:12–21
109. Zervopoulos P, Palaskas T (2011) Applying quality-driven, efficiency-adjusted DEA (QE-DEA) in the pursuit of high-efficiency–high-quality service units: an input-oriented approach. J Manage Math 22:401–417. https://doi.org/10.1093/imaman/dpr014
110. Zuboff S (2010). Creating value in the age of distributed capitalism. McKinsey Quarterly

Dominic Loske is a Ph.D. student in SCM & logistics at the Universidad Católica San Antonio de Murcia, in cooperation with the Institute for Logistics and Service Management, FOM University of Applied Sciences. His main research interest centers on the efficiency of human-AI collaboration in logistics, as well as on learning curves and autonomy when automating human work. Dominic holds a master of science in SCM and logistics from the FOM University of Applied Sciences Munich.

Matthias Klumpp is working at Georg-August-University of Göttingen, FOM University of Applied Sciences Essen and Fraunhofer Institute for Material Flow and Logistics (IML) Dortmund. He received his Ph.D. from the University of Leipzig, his Habilitation from the University of Duisburg-Essen and was in visiting positions at University of Twente, EUI Florence, University of Lugano (IDSIA), and University of Modena. Addressed research topics are sustainable, digital (AI), and resilient supply chains with a focus on HCI and qualification.

A New Labor Unionism in Digital Taylorism? Explaining the First Cycle of Worker Contention at Amazon Logistics

Riccardo Emilio Chesta

Abstract This chapter analyzes the cycle of worker contention at Amazon logistics in Italy, explaining the link between digitalization in logistics and worker mobilization. In this case, workers' collective action has shown the critical aspects of the Amazon model and its introduction of algorithms in delivery systems. The analysis of the mobilization is twofold. Focused on worker power, repertoire of action and claims, it explains on the one hand, how technology has affected working conditions and on the other hand, how worker organization triggered specific forms of unionism. While Amazon's algorithmic innovation brings in what can be called a digital taylorism, this case of contention shows the forms of worker mobilization, which led to a critique of its implications and to new processes of unionization. Finally, the case disconfirms any determinism and shows how technological innovation can be negotiated as part of collective bargaining where the regulation of an employment relations system plays a central role.

Keywords Digital taylorism · Technology · Work organization · Collective action · Labour unionism · Amazon · Logistics

1 Worker Contention and the Social Shaping of Digital Technologies

As in any other wave of technological change, the growing integration of information and communication technologies (ICTs) in different sectors of the economy have not only challenged existing forms of work organization but also introduced new business models and shaped new forms of employment relations [15, 32]. The recent diffusion of algorithms in particular [18, 26] has not only expanded the opportunity for digital work and e-commerce, but rather modified the relations between worker control and autonomy, as well as reshaped the forms of workers' organization and

---

R. E. Chesta (✉)
Scuola Normale Superiore, Florence, Italy
e-mail: riccardoemilio.chesta@sns.it

M. Klumpp and C. Ruiner (eds.), *Digital Supply Chains and the Human Factor*, Lecture Notes in Logistics,
https://doi.org/10.1007/978-3-030-58430-6_10

mobilization [47, 48, 54, 55]. Digital technologies create new spaces of communication and intermediation between work demand (company) and supply (workers), but also redefine specific relations between them [11, 42, 52, 53]. Studying technology in relation with workers and worker agencies is therefore an important perspective which can avoid deterministic analytical shortcuts by showing effects of technology that are uncertain to designers and stakeholders and the power of social actors to correct and modify its course [35].

Literature has shown how technological innovation, far from taking place in a societal vacuum, is always implemented for specific social interests and in specific socio-economic contexts [33, 34]. In the social sciences, a perspective centered on the 'social shaping' of technology better enables a link between the macro-sociological interconnection between technology and social structure with the meso-sociological level of power relations and social interaction among actors involved in its design [31]. In this sense, innovation driven by ICTs is best analyzed as a process, which takes place in a specific labor market regime, enhancing processes of labor empowerment or fragmentation as well as worker control and autonomy.

At the same time, the disruptive aspect of technological change under certain conditions can trigger new forms of collective action, which takes the form of a sustained set of interaction among workers, defining common interests and coordinating in shaping and regulating new technologies on the job [13]. Moreover, the newness of a technological environment directly affects non-technical dimensions like the organizational model and the agency of the workforce, and it implicitly creates a fertile ground for innovation in worker mobilizations [38, 39].

Literature on labor process theory has shown how innovative technological features in working conditions can be better perceived in the first instance by workers rather than actors outside the workplace. Indeed, many cases in the history of labor show how the routinary experience of workers can to some extent be revelatory aspects of the working activities that are not available to managers or external actors [7, 28]. Such knowledge and practices potentially trigger forms of resistance on the job, like the practice of «games of production» [8], as well as new forms of conflict which target the downsides produced by a specific design and application of technology causing deskilling processes and degradation of work [5]. In this regard, both literature in social movements and labor studies suggest using the importance of worker power and resources to explain the rise of worker mobilization, as well as variations in its outcomes, like the capacity to organize or obtain collective improvements [23, 24, 30, 40, 56].

This chapter tries to explain how workers mobilized in a sector affected by the introduction of digital technologies. It does so by analyzing what scholars in the field of collective action and contentious politics calls the «cycle of contention» [45, 51], focusing on the mobilization of Milanese drivers working at Amazon logistics in Italy. The outcomes show how the specific forms of worker mobilization have only partially targeted digitalization as the main cause of discontent. Indeed, workers contention emerged as an outcome of three conditions: the absence of a system of employment relations, the fragmentation of the labor market and the rigidity of an organizational model which can be described as a "digital Taylorism" [6, 23, 43].

This can be defined as a digital extension of some principles of traditional Taylorism, considered as a process of separation from execution, human work routinization, and automation of work rationalization and monitoring through algorithms, which finally allows the documentation of working performances through human capital metrics and remote control.

This chapter explains the rise of worker mobilization and the implications of digitalization on work and workers capacity to mobilize. It is structured as follows. After a brief methodological note in Sect. 2, it introduces the structure and organizational model of Amazon in Sect. 3, and by focusing on worker power (Sect. 4), it explains the forms of action (Sect. 5), the claims of the mobilization process (Sect. 6) and their relation to unions (Sect. 7). It finally concludes by suggesting some elements of comparison and generalization for future research in the field of labor sociology and digitalization (Sect. 8).

2 Data and Methodology

This research on the worker mobilization at Amazon Logistics in Milan is based on new data collected through an abductive process [46] during an intensive fieldwork conducted with qualitative methodologies, mixing ethnographic observations and interviews in two different moments between April 2019 and September 2019 in the urban area of Milan, Italy. Starting from a puzzling situation of spontaneous worker mobilization in a logistic sector affected by digitalization, this case-oriented research [36, 37] develops new knowledge and hypothesis through a grounded and in-depth inquiry. By looking at the relations between workers and digitalization, this research investigates the way worker power triggers a mobilization process innovating repertories of action, claims and relations to unions. It presents a new case study that can be compared with other sectors affected by similar dynamics of technological innovation and forms of worker mobilization. In order to detect mechanisms and explain the dynamics of mobilization, this case required "robust descriptions" [49, 51] grasping how individuals understand and reconstruct the situation they inhabit to enact the mechanisms, which in this case are the capacity to collectively mobilize, to raise collective issues regarding working conditions, and finally to trigger an organization process affecting both unions and employer associations.

A preliminary analysis of the process has been conducted through media coverage. The newspaper «La Repubblica—Milano» has been used to detect the main information regarding Amazon protest events due to the detailed online coverage of the entire urban area. A corpus of 26 articles between 3.11.2015 and 25.02.2019 has been created concerning the expansion of Amazon and the rise of strikes in the Milanese area. These data have been triangulated with information from the company's annual report[1] and the websites regarding Amazon Italy in general[2] and Amazon Logistics in particular.[3]

The primary sources consist mainly of semi-structured interviews and non-participatory observations at spontaneous worker assemblies. Following a selection process, which triangulated snowball sampling and suggestions by random workers, ten individual semi-structured interviews have been conducted with eight workers belonging to different service providers and playing a key role in the mobilization as well as two representatives of the main union involved. Interviews lasted between 45 min and 8 h 20 min. They have been recorded with the consent of the interviewees, anonymized, transcribed by a research assistant and merged into a text subject to content analysis. The text has been coded into sections that refer to specific variables regarding collective issues (working conditions, forms of organizations, relations to unions, collective claims).

In addition, I conducted interviews with key informants and participated in three worker assemblies. Non-participatory observations were conducted in three different Amazon *stations* near Milan, getting in contact with workers ending their daily shift. Data are based on fieldwork notes taken directly during worker assemblies. In particular, two observations have been recorded with the explicit consent of workers and transcribed.

3 Amazon Logistics and its Expansion

Amazon was a multinational company founded in Seattle as an online bookstore. In two decades the company became an everything store, an almost monopolistic player in e-commerce in Western countries [44], which nowadays globally employs 560,000 workers—called Amazonians in the company's jargon. Its size is visible in its brand value of 315.51 billion dollar, the world's most valuable company followed by Apple (309.5) and Google (309.0).

In fact, Amazon recruited 130,000 new employees only in 2017 [2, 25]. Specifically regarding Italy, the company has invested 1.6 billion euros and created 5,500 jobs. The first Italian distribution center was opened in 2011 in Castel San Giovanni near Piacenza followed in 2015 by a new center for urban distribution at Affori near Milan, and from 2017 on by other centers throughout the country, particularly in the regions of Pedemont, Tuscany, Veneto and Lazio.

The Milanese case is particularly relevant and at the same time exceptional with respect to the other centers in the country due to the innovative environment of the urban area which concentrates one of the highest rates of e-commerce demand. Milan is the central urban area of Lombardia, a region which nowadays covers 3,226 merchant online activities, which is 1/5 of the Italian market for e-commerce.

[1] https://www.annualreports.com/Company/amazoncom-inccompany.

[2] https://www.aboutamazon.it/centri-di-distribuzione/investimenti-locali/investimenti-locali.

[3] https://logistics.amazon.it/.

In particular, 10–15% of parcel deliveries from e-commerce are concentrated in the Milanese urban area,[4] which anticipates most of the trends that progressively extend to the whole country. Data are volatile because of the fragmentation of the supply chain and the Amazon system of subcontracting, where the mortality of small companies operating in the sector is high. However, some estimations calculate that among the workforce employed in the delivery service in Milan about 1,500 workers are employed by companies operating for Amazon.

Even though Amazon is a player which entered in the field of Italian logistics quite recently, its model proposes at the same time an innovative model in the systems of delivery and distribution, emphasizing cost cutting strategies and customer orientation. This innovation thus necessitates two conditions: on the one hand, more flexibility in the employment regime of the workforce, and on the other hand, a more rigid control on the pace of work and workloads.

In Milan, Amazon opened a new warehouse in November 2015 near Affori—located in the Northern hinterland—dedicated to the Amazon Prime Now service, promising a service of delivery in just one hour, 24/24 h, 7/7 days (La Repubblica—Milano, 3.11.15). The company builds in 2016 with another plant in Porta di Mare—near Rogoredo, south of Milan.

Between 2012 and 2016, Amazon Italy increased the number of employees, opened another warehouse near Origgio—located between Varese and Milan—and in May 2018 in Buccinasco—in the outskirts of Southern Milan. The position at the borders of the city and the availability of a new workforce from all over the hinterland of Milan allow Amazon to operate in key sites for the concentration and distribution of commodities.

Amazon recruits drivers—a contractual definition that, as we will see, is already subject to contention—through societies or consortia, which manage delivery services as service providers or subcontractors. Give the volatility of Amazon's volumes of delivery and the scarce experience of service providers in dealing with it, many companies in turn make use of subcontracts, broadening a pyramidal system that sees Amazon as the head of the supply chain.

The new Amazon workforce, which operated at the new *stations* in the Milanese area from 2015 on, had no previous experience of the Amazon model in terms of work organization and regime of employment. Consequently, unionism and worker organizations are absent at Amazon stations. In this context, the introduction of algorithms and the absence of a collective bargaining regime are the two conditions that triggered a cycle of strikes that emerged since May 2017 and led to a unionization of a significant part of the workforce (1/3 of 1,500 Amazon drivers).

In the process of expansion, which in the last two years made Amazon one of the first operators in logistics of e-commerce in Italy, the analysis of the rise of worker mobilizations in a multinational company, which has not been permeable to unions,[5]

---

[4]https://www.regione.lombardia.it/wps/portal/istituzionale/HP/lombardia-notizie/DettaglioNews/2018/06-giugno/18-24/e-commerce-imprese/e-commerce-imprese.

[5]From a comparative perspective, Amazon arrives in Europa already in 1999 when it settles in Germany with the warehouse in Bad Hersfeld which remains the only one until 2006. The company

allows to formulate some preliminary hypothese about the possibility for workers to regulate a sector disrupted by the arrival of one of the most innovative global players.

4 Worker Power and Organizational Mode

The employment regime in which an Amazon driver operates is characterized by some specificities shared with other companies in logistics, which usually recruit through subcontracting. In this case, the drivers are not employed directly by Amazon but through the intermediation of subcontractors which sign an agreement with the Italian section of the Seattle multinational company, which is in this case the "head" of the supply chain. Subcontractors operate at specific Amazon stations where parcels are stocked and distributed, depending on quantity and typology. This means that, although workers start their daily work at the same relative station, they work for several companies which provide for Amazon's different services and put in place different contracts. The fragmentation of work is then due to these conditions: multiplication of subcontractors due to service differentiation and consequently differentiation of contractual standards. As a union representative describes, fragmentation regards workers as well as service providers: When we first penetrated in the sector, it was a mess: trade contract, agri-food contract, postal or agroindustry contract. When we've arrived, we started claiming the logistic contract, since Amazon was already applying that contract to warehouse employees" (INT7) (Fig. 1).

Left to the discretion of service providers, contractual fragmentation was mainly a short-term strategy of adaptation to Amazon's conditions by trying to find the best framework guaranteeing low costs of work. However, although service providers

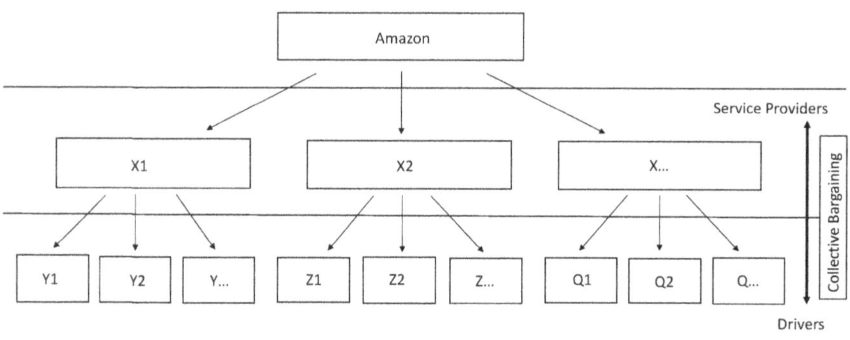

Fig. 1 Employment relations among Amazon logistics, service providers and drivers

plans afterwards a new expansion opening another nine plants in Germany. Also for geographic reasons, Germany is indeed the main logistic heart of Europe. Notwithstanding, Amazon is a company without unions until 2011 when the German union for the service sector Ver.di initiates a mobilization process and tries to unionize workers with the main goal to propose the recognition of the collective contract of trade [4, 10].

contributed to fragmentation, Amazon's protocols regarding work organization and technologies tend on the contrary to unify working conditions.

Indeed, while the contractual regime is an aspect managed by service providers, the most salient aspects of work organization—the algorithmic routes—are designed by Amazon management. In this regard, drivers follow principles that are similarly applied at Amazon warehouses driven by algorithmic automation [29] and in which digital monitoring extends beyond traditional forms of supervision based on physical and spatial copresence.

In this regard, the workplace level is the key starting point to understand how worker started to recognize their common condition and to organize collectively.

The daily shift starts when workers arrive in the morning at their specific Amazon station and they are assigned to specific stops fluctuating between 100 and 120 per day, depending on the route and difficulty (weight, size, and typology). After loads of delivery are assigned by a dispatcher employed by the service providers, workers gather parcels in a van provided by their respective company. With a mobile-phone they log on to their account through a specific Rabbit-Amazon app which assigns routes and keeps track of the parcel delivery and van travels. Although theoretically every subcontractor is free to organize its service, some principles are fixed by Amazon, like the tracking of the deliveries (with all the customer data included), the timing of deliveries (linked to the customer service) and the use of an algorithm to calculate them. Workers have to deliver their number of assigned packages in a working day composed of 9 h and 15 min with a 30 min break included. The algorithms that determine their routes and workloads calculate every shift in terms of expected number of packages per time of delivery and distance, constantly updating these measures depending on efficiency criteria and worker performances.

Algorithms are designed and updated by Amazon, which has therefore not only the power to decide workloads and paces of work, but also finally owns the data produced by the drivers. This ultimately happens with no obligation from Amazon to negotiate its conditions with other parties—subcontractors and workers. This is how the company officially declares the condition of subscription to their subcontractors:

> Amazon stipulates its contracts with their delivery service providers. Service providers employ their own drivers and are the only responsible for their personnel. The personnel of the provider cannot participate to any plan for benefits or other benefits available to other Amazon's employees. Delivery service providers and their relative personnel do not have the authority to oblige Amazon or other affiliates to sign any contract or obligation. (As you can imagine, this reply has been written by our lawyers ...).[6]

In this framework, the construction of a common platform for negotiations that reduces the fragmentation among workers and different service providers is thus an outcome of a process of collective aggregation triggered by workers as this union representative states:

[6] From the section "Frequently Asked Questions", to the relative question "What are the juridical relations between Amazon and the drivers?", from Amazon's official website: https://logistics.amazon.it/.

> We proposed the national contract of logistics and, by strikes and negotiations, we convinced different companies to sign the same agreement. At the beginning, they were only few, 4 companies, but then 5, 6, 7, 8, 9... at the hand they decided to establish an employer association: we signed the first supply chain agreement on 11 October 2018. From this moment, everybody complied with that. (INT7)

This process of aggregation led to acknowledgement of the core issue at stake: Amazon's organizational model. As it is expressed by a worker participating in the strikes:

> The core of the issue is the organization of work at Amazon. Within this, there's Assoespressi which is the interest group of the companies [service providers] that are trying to mediate. But this can be controversial because we want to negotiate with Amazon. They [service providers] told us to save the routings that were not matching... that algorithms would have recalculated that routings. But algorithms recalculated only if we were finishing our stops earlier. I had shifts that lasted 12 h and the algorithm never recalculated anything. (INT5).

In this regard, algorithms are the key aspect of Amazon's model. Algorithms are encoded procedures that transform input data into a desired output based on specific calculations [18, 26]. Their extension to a new working environment rises new questions about their implications for human skills as well as the relation between autonomy and control [48]. Applied to delivery workers in logistics, algorithms are translated in digital tools to automatically monitor worker performance and to eliminate delays and slowdowns in the process of delivery, with the power to localize workers and access data linked to their activity. The calculation of workloads and working times through algorithms frequently does not take into account problems like urban traffic or work-in-progress on the way. Moreover, algorithms are still not able to classify the site of delivery or customers with specific needs (e.g., apartment building in a condominium with a porter or not, presence or absence of a front door, elevator, and so on) and consequently the precise variation in times of delivery. Thus, the difference is still made by the «human factor» of workers who, knowing the best alternative way to avoid traffic, or the typology of a building where to deliver a set of parcels, have a specific expertise that algorithms still donot have.

The scientific application of algorithms in the management of routes and deliveries can be defined as a specific extension of «digital Taylorism» in logistics, as a prosecution through digitalization of classic organizational models of scientific management in physically defined working spaces [1]. In particular, digitalization allows the pursuit of «one best way» in the delivery process with the calculation and forecasting of pace of work, and a permanent monitoring of deviations and dead time without the physical co-presence of a supervisor. More specifically, with the digital monitoring for performances, the management can automatically record and extract data produced by workers like strategies of route adjustment or knowledge of the territory.

Thus, this attempt to scientifically model an "assembly line in the head" [47], that is at the core of the claims emerged in the worker mobilization, has a twofold meaning. It leads to a rigidity of the model and potential deskilling, but also to controversies regarding the ownership and the value of workers' data and knowledge. As it is explained in the next sections, this is the main reason for the mobilization and one of the main claims that become a core argument for the unions.

5 Mobilization Process and Forms of Collective Action

The first mobilizations of Amazon drivers in 2017 was triggered by a spontaneous group of workers with no previous union membership or experience. While trying to build alliances among workers employed by different companies, this process attracted traditional trade unions which, while being pushed toward more contentious repertoires of action, immediately provided organizational, legal and technical support. The forms of struggle adopted reflect the traditional repertoire of worker movements, like strikes and pickets out of the Amazon stations and van deposits. Among all the repertoires of action, picket lines are the clearest expression of worker power, being one of the most effective means that directly cause the stop of deliveries, attract workers, and oblige service providers to open a negotiation. Indirectly, the resonance of its effects attracts the attention of the media and thus potentially contributes to attract alliances with other parts of society.

The first Milanese strike at Amazon Logistics took place in May 2017, with a picket lines of workers in front of the depot in Origgio, a crucial site where packages get concentrated and prepared to be distributed in all the storages of the country. Workers denounce the presence of "killing time schedules and illegal work" (La Repubblica—Milano, 11.05.17). While declaring that the company imposes unsustainable workloads reaching 14–15 h a day, they at the same time placed emphasis on the use of para-illegal and illegal working times to attain shifts which are unsustainable for service providers and workers. Indeed, this situation reflects typical dynamics of complex supply chain structures in other logistic sectors, where illegal companies or bogus cooperatives are often the norm [3]. The impossibility to adapt to the flows of Amazon loads, which constantly expand because of continually emerging new programs like Amazon Prime Now, leads service providers to make use of black work or use rhetoric which tend to mystify employment relations with workers. As one of the main participants of the strikes declares:

> They [the service providers for Amazon] tell us: we are a start-up. We are new, let's help each other. They keep telling us that we're a start-up. Still now, after two years and a half, they keep saying this. The route is paid [by Amazon] about 200–210 Euro. The service provider pays the worker, the van, the gasoline and then it earns the rest. This system is still in progress, they can't really fit into it with that kind of payment. [...] So, they start with these magic tricks to survive [INT3].

In this framework, threefold are the claims advanced by workers: "Respect of the national contract of logistic", the reduction of the fragmentation of the supply chain and "more clarity concerning payments, both for hours of overtime and for bank transfers and salaries" (La Repubblica—Milano, 11.05.17).

These actions, which started spontaneously from about 70 workers, triggered a new attempt of "first agreement in the belly of the giant" as the trade unions later commented—and particularly the main interlocutor of the Milanese drivers, the FILT-CGIL (La Repubblica—Milano, 12.05.17).

On June 27, another protest event regarding the warehouse in Affori and Origgio occured, where the drivers of different subcontractors criticized the new contractual framework which considers them postal workers, misrecognizing both their skills and reducing the costs of work (La Repubblica—Milano, 28.06.17).

A first direct outcome of the strike was the declaration by the the subcontractors to be open to negotiations. Any exchange between workers and these companies—like RPost, SailPost, Helpe, Madilo, Sum, among others—occured without the participation of Amazon management, which never takes part in negotiations. In parallel, new protest events spread in other Amazon poles involving other workers, like warehouse operators during the «Black Friday» in November 2017 in Castel San Giovanni near Piacenza.

Amazon drivers struck again in September 2018 in Origgio, Affori and the new Amazon station in Buccinasco (La Repubblica—Milano, 14.09.19), which on 18 October 2018 leads to the first "supply chain agreement" signed among the employees associations (Assoespressi and Fedit), the union representatives of FILT-CGIL. FIT-CISL and UIL Transport Lombardia, which applies the national contract of logistics and puts a limit to the daily working time (Fig. 2).

With the opening of the negotiation process involving the relative subcontractors, the wave of strikes and their claims grew in numbers and in the generality of claims [19, 41], reaching an important consensus among workers. The cycle of protest symbolically consolidated on 24 February 2019 with the declaration of the first regional general strike of Amazon workers in Lombardia, an event which culminated with the presence of Maurizio Landini, just elected new general secretary of the main

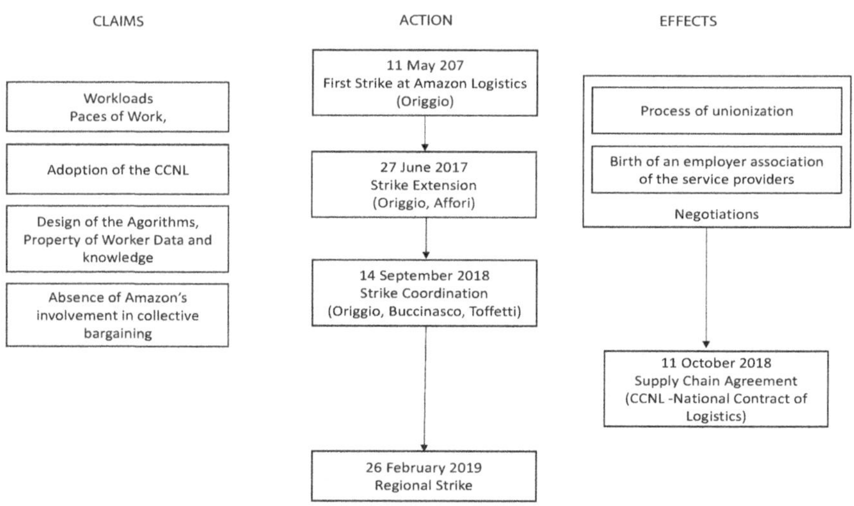

Fig. 2 The first cycle of contention at Amazon Logistics

Italian union, the CGIL, at the end of the demonstration. (La Repubblica—Milano, 25.02.19).

Concluding, while the elements of novelty of these strikes regard the main e-commerce multinational company which is disrupting the logistic sector with an innovative organization based on algorithms, the mobilization process and the wave of strikes threw a new workforce into the arena which pushes the unions to adopt repertoires and mixes institutional and social movement tactics.

The conflict reflects a traditional process of labor mobilization against a multinational company but without windows of opportunity for a direct negotiation with the main contractor. In this sense, the outcomes of the spontaneous worker mobilization have been, firstly, to attract the support of traditional unions—mainly the FILT-CGIL—which was able to provide resources and expertise to workers to be effective and to initiate the first negotiations. Secondly, this opened an intermediary level of negotiation which led to the constitution of an employer association. Finally, if these dynamics of mobilization are quite common in new working sectors characterized by the absence of clear regulations, the main novelty of the worker mobilization will be found indeed in the claims, which redefine the nature of work and its relation becomes algorithms.

6 Claims of Mobilization

As it becomes clear by looking at the main protest events, all along this first cycle of strikes, algorithms were at the core of workers' claims. Indeed, pace or loads of work are directly determined by algorithms designed by Amazon without any negotiation with unions or service providers. Algorithms renew a traditional problem regarding how technology reshapes work control and autonomy but at the same time extend the problem to the property of data and knowledge, the human factor and the soft skills. Milanese drivers dispose of skills which do not easily fit to the guidelines elaborated by the algorithms. Algorithms do not contribute only in reducing time and facilitating routes of delivery, but they also contribute in lowering the professional status of workers, which consequently should adapt and execute what is codified by algorithmic prescriptions. This is an element which became a key claim of the official discourse of the union which supported the organization of the strikes:

> The reason why we are strong here is the professional aspect of the workers. There are workers strongly aware of their know-how, an identity element that exists *before the union* and that *they ask the union* to represent. This is the point of the dispute opened with Amazon's algorithm. The point is that workers know how to do a job whereas there's an algorithm which in the best scenario tells them what to do and in the worse steal them the job. In the first case, the algorithm doesn't know the territory – if there's a porter in a building who facilitates the delivery process, if there's an old lady who can open the door or not and so on. That is the human component which the worker is able to acknowledge. This relation with the territory and with the customers is not kept into consideration by the algorithm. This worsen the quality of work and service. In the worst case, the algorithm steals the job because it incorporates a series of data while the workers are following a new path, a quicker one

> because it is linked to his direct experience. Therefore, the worker is giving information and data to the algorithm which is recording them. In this way, while processing this information, the algorithm takes possession not only of the knowledge of the worker but it makes him substitutable, because it can be substituted in the future, maybe by a robot, but simply with another worker. This is all professional skill which is not paid. (INT1)

The problems described above show the importance of what labor scholars call «working knowledge» [28], that is the practical expertise developed on the job. In the case of Milanese drivers, this knowledge of the city and the territory is at the core of their professional skills and the necessary *know-how* to perform the delivery service. Algorithms are therefore criticized not primarily as a tool of surveillance, but as a source of workload increase which accelerates pace of work and produces new work-related risks, while above all expropriate skills (worker knowledge). Therefore, claims target company exploitation of digital tools that, behind the aura of a technological innovation (overcoming physical intermediation, accelerating services and creating new jobs opportunities), mainly fragmentate labor and elude worker rights like their professional recognition.

Proposals elaborated by workers are mainly twofold, trying to strategically link the quality of work and service. Firstly, to "bargain the algorithm"—using an expression that became official among the unions [20]—a preliminary recognition of the skills of the workers is necessary. Secondly, the «delivery routes» defined by algorithms need to be negotiated depending on more sustainable criteria which deal with working conditions that are at the same time embedded in the concrete context of the city. In this sense, the opening of new Amazon stations should be taken into account in programs of urban planning which read the phenomenon in its complex relation with the territory.

Consequently, worker claims suggest the search for new alliances between citizens and customers in order to be successful, which is in line with a contemporary tendency of labor politics to seek support and to build coalition with civic organizations and communities [12, 14]. If the quality of the service is indeed inversely proportional to the rapidity of the delivery, the issue of pace of work can be negotiated. However, if the claims suggest important openness toward an extension of the actors involved in the issue, at the moment no real alliance will be visible among workers and consumer organizations and the cycle of protest will carry the characteristics of a traditional labor mobilization [9].

7 Worker Mobilization Between Old and New Unionism

The most visible outcome of the mobilization of Amazon drivers has been the revitalization of traditional unions, which have entered a field where collective bargaining was almost absent [17]. Concretely, this is visible not only in protest events but also in the establishment of the first worker assemblies organized out of the Amazon stations—neither Amazon logistics nor subcontractors provide spaces for worker associations. As previously explained in Sect. 4, this should be a duty that concerns service providers, which officially allow the right to associate but do not have a physical space dedicated to worker associations, which therefore take place in various public spaces nearby (principally surrounding parking spaces). One of the key aspects of the mobilization is how groups of workers without any specific experience in previously in the chapter "Amazon Logistic" or in organizing started to see the union as an important platform to raise their voice concerning specific needs and to organize and attain collective goals.

This needs to be explained against three negative conditions: First, collective action takes place in a field characterized by precarious and unstable contracts, where costs of mobilization are high and that theoretically should not leave too much space for conflict and protest. Second, the working activity of drivers is an individualistic task which performance is determined by a dispatcher, and automatically localized and surveilled with Amazon's apps installed on tokens when given to the workers, and where they log-on through a company account at the beginning of their shift. Third, their working experience, beyond contractual fragmentation, leaves little space for collective gatherings, being limited to the collection of orders in the morning and the return to the station at the end of the working day.

Beyond these conditions which theoretically constrain the opportunity to develop a shared set of complaints and a collective strategy to raise them, the mobilization has grown over time and reached a level of collective organization which brought many workers to unionize, spreading beyond the single specific conditions determined by the different subcontracting companies. As these two drivers state:

> I entered for the first time with the FILT (The Italian Federation of Transport Workers, CGIL) because before I didn't trust unions that much. I would even say, quite the contrary. Today I think that the unions do not have the reputation they should. They are perceived as a political part, thing that is not right if you have to protect a worker. It's right that you have your ideas, but out of the schools, at the park, at the mall, when you talk about the unions, people relate like to a political party. They do not relate to an institution of the worker. This demonstrates that in Italy companies do whatever they want. I refer to what my grandpa did, the war, the partisans of the Italian resistance. They thought, they have done the "iron age" with the Magneti Marelli. There you really had the unions, what should come back today, above all to govern these multinationals which think only and exclusively to their profit. (INT5)

> Before, I was working for a delivery company, then I worked as a food-delivery, for a platform, but it didn't work well. I addressed my concerns to the unions, to solve these problems and then I met an old colleague with whom I worked for the first delivery company. He sent me to a union representative of the CGIL which told me he needed a representative in a station and I accepted. (INT2)

These two testimonies show the emergence of a unionism that is defined by the capacity to give direct answers to practical needs and specific problems, in opposition to a traditional alignment which sees unionism as part of a broader political framework. The grassroots unionism typology can be seen in the definition of «institution of workers for workers».

The support of structured unions, as the FILT-CGIL, has been central, although, as suggested by literature on resource mobilization [30], workers already disposed of power and resources at the level of the workplace. Complaints have been easily expressed through spontaneous strikes since workers had the power to block vans at

Amazon stations, while the brand of the company implicitly helped workers to easily reach a public visibility and gain media attention. However, despite these conditions, the economic and symbolic costs of mobilization were not sustainable in the long-term without an elaborated strategy capable of negatively affecting the company while reducing risks for the workers (layoff, economic or legal sanctions). At the same time, the structure of workloads (9 h 15 shifts on five days a week) encourages the intervention of a structured organization with professional unionists rather than self-organized workers. Workers did not dispose of enough time, resources and experience to organize spontaneously a collective bargaining. Among the workers mostly were adults in need of a stable job and there was a considerable number of migrants, which had no experience of activism, in need of an external support to become socialized to the mechanisms of collective bargaining.

Finally, the mobilization of drivers has shown that institutional and conflictual repertoires are not two exclusionary categories, but that they both can provide resources for the revitalization of unionism [39]. More specifically, the main issue at stake is how this localized and sectorial mobilization can participate in the near future to a process of generalization that constructs new alliances with other sectors of labor in digital capitalism.

8 Conclusion

By looking at the relations between work, organization and technology, this chapter has explained the causes that led to the first cycle of worker mobilization in the delivery sector of Amazon logistics as well as the dynamics of mobilizations that established the first forms of collective bargaining at one of the most innovative multinational players in the e-commerce. This case study has shown how digitalization has been only partly the cause of mobilization. The introduction of algorithmic control has not been targeted and criticized in itself, but as a tool put in place by the head of the supply chain which, in a deregulated labor market, bypassed collective bargaining. In doing so innovation favored rigidity of worker control and flexibility only on the side of employment regime, finally triggering processes of deskilling and work degradation.

In line with the organizational model introduced at Amazon warehouses, the introduction of algorithms favored a digital Taylorism taking the form of «an extended assembly line in the head» of drivers [43, 47], automatically separating conception from execution, extending monitoring and extracting data from worker performance through digital technologies. In terms of interaction dynamics, the effect of the mobilization process has been threefold. First, it has been the rise of a collective identity among workers which Amazon did not recognize, innovating the field of traditional unionism and pushing unions to adopt more contentious tactics. Secondly, it has pushed service providers to organize in an employer association. Positive outcomes of the mobilization process have been the adoption of the national contract of logistics, but also the introduction of a preliminary system of employment relations.

At the same time, this success has been partial. If Amazon logistics is now still able to bypass collective bargaining, worker claims cannot affect the core of the problem. Given the difficulty of the service providers in dealing with the high volatility of deliveries linked to Amazon e-commerce services, an integrated system of collective bargaining would therefore not only be beneficial to workers but also to the same companies providing delivery services for Amazon, and Amazon as well. Indeed, the recognition of a collective contract for the drivers will be beneficial to Amazon, avoiding firstly the risks linked to the volatility and fragmentation of the workforce, and secondly enhancing the effectiveness of a uniform process of collective bargaining that can provide more quality to work and services, while also improving the reputation of the company.

Appendix and Sources

Media Coverage

Articles on the Newspaper «La Repubblica—Milano» Regarding Amazon Between 3.11.2015 and 25.02.2019

Interviews

I1. Interview with Regional Secretary—FILT CGIL Lombardia, Milan, 13 March, 2019.
I2. Interview with Amazon Driver—Milan, 18 March, 2019.
I3. Interview with Amazon Driver—Milan, 18, March, 2019.
I4. Interview with Amazon Driver—Milan, 18 March, 2019.
I5. Interview with Amazon Driver—Milan, 18 March, 2019.
I6. Interview with Amazon Driver—Milan, 15 April, 2019.
I7. Interview with Trade Unionist—FILT CGIL, Milan, 27 September, 2019.
I8. Interview with Amazon Driver—Milan, 28 September, 2019.
I9. Interview with Amazon Driver—Milan 29 September, 2019.
I10. Interview with Amazon Drivers—Milan, 29 September, 2019.

Non-participant Observations

A1. Participation at Worker Assembly Buccinasco, Milan, 10 April, 2019.
A2. Participation at Worker Assembly Toffetti, Milan, 9 April, 2019.
A3. Participation at Worker Assembly Via Lombroso, Milan, 15 April, 2019.

References

1. Bain P, Watson A, Mulvey G, Taylor P, Gall G (2002) Taylorism, targets and the pursuit of quantity and quality by call centre management. New Technol, Work Employ 17:3
2. Berg N, Knights M (2019) Amazon how the world's most relentless retailer will continue to revolutionize commerce. Kogan Page, London
3. Bologna S, Curi S (2019) Relazioni industriali e servizi di logistica. Uno Studio Preliminare, in Giornale Di Diritto Del Lavoro E Di Relazioni Industriali 161(1):125–156
4. Boewe J, Schulten J (2017) The long struggle of the Amazon employees. Rosa Luxemburg-Stiftung, Brussels
5. Braverman H (1974) Labor and monopoly capital. The degradation of work in the twentieth century. Monthly Review Press, New York and London
6. Brown P, Lauder H, Ashton D (2011) Digital taylorism, in The global auction. The broken promises of education, jobs and incomes. Oxford University Press, Oxford
7. Bruland T (1985) Industrial conflict as a source of technical innovation: the development of the automatic spinning mule. In: McKenzie D, Wajcman J (eds) The social shaping of technology how the refrigerator got its hum. Open University Press, Philadelphia
8. Burawoy M (1979) Manufacturing consent. Changes in the labor process under monopoly capitalism. The University of Chicago Press, Chicago
9. Caruso L, Chesta RE, Cini L (2019) Le Nuove Mobilitazioni Dei Lavoratori Nel Capitalismo Digitale:Una Comparazione Tra I Ciclo-Fattorini Della Consegna Di Cibo E I Corrieri Di Amazon Nel Caso Italiano in Economia E Società Regionale 1(19):61–79
10. Cattero B, Onofri, V (2018) Orfani delle istituzioni. Lavoratori, sindacati e le "fabbriche terziarie digitalizzate" di Amazon. Quaderni Rassegna Sindacale—Lavori 1:7–28
11. De Stefano V (2015) The rise of the just-in-time workforce: on-demand work. Crowdwork Labor Protect Gig-Econ Comparat Labor Law Policy J 471–501
12. Dorigatti L (2015) Sindacati e Alleanze Sociali. Riflessioni a Partire dalla Campagna «Fight for $15», Quaderni di Rassegna Sindacale 3:125
13. Edward P, Ramirez P (2016) When should workers embrace or resist new technology? New Technol, Work Employ 31(2):99–113
14. Fine J (2005) Community unions and the revival of the American labor movement. Politics Soc 33(1):153–199
15. Fischer G, Downey G (2006) Frontiers of capital. Ethnographic reflections on the new economy. Duke University Press, Durham and London
16. Frege C, Heery E, Turner L (2003) Comparative coalition-building and the revitalization of the labor movement. Presented at: 55th international industrial relations association annual meeting. Washington, D.C., 3–5 January
17. Frege C, Kelly J (2004) Varieties of unionism: strategies for union revitalization in a global economy. Oxford University Press, Oxford
18. Gillespie T (2016) Algorithm. In: Peters B (ed) Digital keywords: a vocabulary of information society and culture. Princeton, Princeton University Press
19. Givan RK, Roberts K, Soule SA (2010) The diffusion of social movements: actors, mechanisms, and political effects. Cambridge University Press, New York and Cambridge
20. Gramolati A, Sateriale G (2019) Contrattare l'Innovazione Digitale. Una Cassetta degli Attrezzi 4.0. Ediesse, Roma
21. Glover J (2013) Digital taylorism: hybrid knowledge professionals in the UK ICT Sector. In: Evans C, Homes L (eds) Re-tayloring management. Scientific management a century on. Routledge, London
22. Grote JR, Wagemann C (eds) (2018) Social movements and organised labour: passions and interests. Routledge, London
23. Lehndorff S, Dribbusch H, Schulten T (2017) European trade unions in a time of crises—an overview. ETUI, Brussels
24. Levesque C, Murray G (2010) Understanding union power: resources and capabilities for renewing union capacity. Transfer 16(3):333–350

25. Kahn LM (2018) Amazon—an infrastructure service and its challenge to current antitrust law. In: Moore M, Tambini D (eds) Digital dominance. The power of Google, Amazon, Facebook and Apple. Oxford University Press, Oxford
26. Kitchin R (2017) Thinking critically about and researching algorithms in information. Communication and Society 20:1
27. Knights D, Willmott H (eds) (1990) Labour process theory. MacMillan, London
28. Kusterer K (1978) Know how on the job. Westview, Greenport
29. Massimo FS (2019) Spettri del Taylorismo. Lavoro e organizzazione nei centri logistici di Amazon». Quaderni di Rassegna Sindacale 3:85–102
30. McCarthy J, Zald M (1977) Resource mobilisation and social movements. Partial Theory American J Sociol 82(6):1212–1241
31. McKenzie D, Wajcman J (1985) The social shaping of technology how the refrigerator got its hum. Open University Press, Philadelphia
32. Moore M, Tambini D (eds) (2018) Digital dominance. The Power of Google, Amazon, Facebook and Apple, Oxford University Press, Oxford
33. Noble D (1979) Social choice in machine design. In: Zimbalist A (ed) Case studies on the labor process. Monthly Review Press, New York
34. Noble D (1986) Forces of production a social history of industrial automation. Oxford University Press, Oxford
35. Orlikowksi WJ (1992) The duality of technology: rethinking the concept of technology in organizations. Organization Science 3:398–427
36. Ragin C (2014) The comparative method. Moving beyond quantitative and qualitative methods. University of California Press, Oakland
37. Ragin C, Becker H (eds) (1992) What is a case? Exploring the foundations of social inquiry. Cambridge University Press, Cambridge
38. Robinson I (2000) Neoliberal restructuring and US Unions: toward social movement Unionism? Critical Sociol 26:109–138
39. Ross S (2007) Varieties of social Unionism: towards a framework for comparison, in just labour. Canadian J Work Soc 11:12–34
40. Silver B (2003) Forces of labor: workers' movement and globalization since 1870. Cambridge University Press, New York
41. Soule S (2004) Diffusion processes within and across movements. In: Snow DA, Soule SA, Kriesi H (eds) The Blackwell companion to social movements. Blackwell, Malden and Oxford
42. Srnicek N (2017) Platform capitalism. Polity, Cambridge
43. Staab P, Nachtwey O (2016) Market and labour control in digital capitalism, TripleC 14(2):457–474. https://www.triple-c.at
44. Stone B (2019) The everything store. Jeff Bezos and the age of Amazon. Little Brown and Company, Boston
45. Tarrow S (1994) Power in movement. Social movements and contentious politics. Cambridge University Press, New York
46. Tavory I, Timmermans S (2014) Abductive analysis. theorizing qualitative research. The University of Chicago Press, Chicago
47. Taylor P, Baine P (1999) An assembly line in the head: work and employee in the call centre. Ind Relat J 30(2):101–117
48. Taylor P, Mulvey G, Hyman J, Bain P (2002) Work organization, control and the experience of work in call centres in work. Employ Soc 16:1
49. Tilly C (1999) Conclusions: from interactions to outcomes in social movements. In: Giugni M, McAdam D, Tilly C (eds) How movements matter: theoretical and comparative studies on the consequences of social movements. University of Minnesota Press, Minneapolis
50. Tilly C (2007) Afterword political ethnography as art and science. In: Joseph L, Mahler M, Auyero J (eds) New perspectives in political ethnography. Springer, New York
51. Tilly C, Tarrow S (2007) Contentious politics. Paradigm Publishers, Boulder-London
52. Vallas SP (2019) Platform capitalism: what's at stake for workers? New Labor Forum 28(1):48–59

53. Van Dijck J, Poell T, De Waal M (2018) The platform society. Public values in connective world. Oxford University Press, Oxford
54. Woodcock J (2017) Working the phone. Control and resistance in the call center. London Press, Pluto
55. Woodcock J, Graham M (2020) The gig economy. A critical introduction. Polity Press, Cambridge
56. Wright EO (2000) Working-class power, capitalist-class interests, and class compromise. Am J Sociol 105(4):957–1002

Riccardo Emilio Chesta is a postdoctoral research fellow at the Carlo Azeglio Ciampi Institute, Scuola Normale Superiore (Florence). He holds a M.A. in social sciences from the EHESS and ENS Paris, and a Ph.D. in political and social sciences from the European University Institute in Florence. He has been teaching assistant in sociological theory at the University of Trento, and visiting scholar at the Urban Democracy Lab, New York University. Focusing on social theory, technology and collective action, his work has investigated the dilemmas of expertise and activism on large-infrastructural projects, industrial plants and the environment. He is currently working on digital capitalism, Industry 4.0 and their consequences for worker mobilizations.

The manufacturer's authorised representative in the EU is Springer Nature Customer Service Centre GmbH, Europaplatz 3, 69115 Heidelberg, Germany. If you have any concerns regarding our products, please contact ProductSafety@springernature.com

Printed and bound by CPI Group (UK) Ltd, Croydon, CR0 4YY
28/09/2026
02224089-0002